Ecological Studies

Analysis and Synthesis

Volume 252

Ecological Studies is Springer's premier book series treating all aspects of ecology. These volumes, either authored or edited collections, appear several times each year. They are intended to analyze and synthesize our understanding of natural and managed ecosystems and their constituent organisms and resources at different scales from the biosphere to communities, populations, individual organisms and molecular interactions. Many volumes constitute case studies illustrating and synthesizing ecological principles for an intended audience of scientists, students, environmental managers and policy experts. Recent volumes address biodiversity, global change, landscape ecology, air pollution, ecosystem analysis, microbial ecology, ecophysiology and molecular ecology.

Valentí Rull

Origin and Evolution of Caribbean Mangroves

A Time-Continuum Ecological Approach

Springer

Valentí Rull
Botanic Institute of Barcelona
Spanish National Research Council (CSIC)
Barcelona, Spain

Institut Català de Paleontologia Miquel Crusafont
Universitat Autònoma de Barcelona
Barcelona, Spain

ISSN 0070-8356 ISSN 2196-971X (electronic)
Ecological Studies
ISBN 978-3-031-57611-9 ISBN 978-3-031-57612-6 (eBook)
https://doi.org/10.1007/978-3-031-57612-6

This Springer imprint is published by the registered company Springer Nature Switzerland AG
The registered company address is: Gewerbestrasse 11, 6330 Cham, Switzerland

Acknowledgments

The main ideas disclosed in this book were conceived during my stay (1990s) in the Venezuelan state oil company PDVSA as exploration biostratigrapher. PDVSA offered an inspiring work atmosphere and authorized the dissemination of findings from various studies, contributing to the formulation and arrangement of ideas that were further developed upon my return to Europe in 2002. While it is impossible to enumerate all the PDVSA personnel who contributed to and backed my research, notable paleontologists Estela Di Giacomo and María Antonieta Lorente merit particular acknowledgment. Close colleagues in the biostratigraphical team were Saileth Cortez, Isbelia Durán, Armando Fasola, Francia Galea, Nancy Hambalek, Mercedes Hidalgo, Freddy Lazarde, Diego Morillo, Rosina Pittelli, Oscar Quintero, Heriberto Rodríguez, Maibí Ruiz, Carlos de Sousa, Irene Truskowski, Miguel Velásquez, and Hender Villalobos. I would also like to acknowledge other less related but influencing partners, such as Juan Francisco Arminio, Fernando Cassani, Omar Colmenares, Jason Crux, Helena Gamero, Santosh Ghosh, Mari Carmen Gómez, Javier Helenes, Roger Higgs, Xiomara Márquez, Selena Mederos, Geoff Norris, Pieter Pestman, Claude Poumot, and Joachim Schiller. My Quaternary skills were obtained with the help of Maximiliano Bezada, Milagro Rinaldi, Maria Léa Salgado-Labouriau, and Carlos Schubert, from the Venezuelan Institute of Scientific Research (IVIC). I would also like to acknowledge some relevant scientists who influenced my paleoecological and evolutionary work in the Neotropics using palynology; these are Alan Graham, Jan Muller, and Thomas Van der Hammen. We stand on the shoulders of giants. No external funding was received specifically for assembling the materials, writing, and publishing this book, which was completed in the Spanish Council for Scientific Research (CSIC), while working in research projects about other topics. The thorough and detailed review of the manuscript by Ernesto Medina is greatly appreciated. As in my previous Springer books, I have received the constant support of the editorial team led by Lars Koerner, as well as the members of the production team, this time represented by Bibhuti Sharma.

Contents

Introduction

1

Abstract

Mangroves possess both intrinsic, ecological, and practical value. Initially, these coastal mangrove forests, situated in the intertidal zone, play a crucial role in safeguarding and sustaining the diversity and ecological dynamics of tropical/subtropical coastal ecosystems. Additionally, they offer a wide array of ecological and cultural services to human societies. Moreover, by sequestering carbon in their organic-rich sediments, mangroves stand out as the foremost blue-carbon ecosystems, contributing to the global warming mitigation. Despite their significance, mangroves face severe threats, ranking among the world's most endangered ecosystems. Sustaining current deforestation rates poses a substantial risk to their long-term survival. Consequently, this predicament has spurred the initiation of numerous global initiatives aimed at conserving and restoring mangroves, necessitating a foundation of robust ecological and evolutionary knowledge. This book is focused on mangroves from the Neotropical Caribbean region and aims to provide a comprehensive ecological and evolutionary view of these ecosystems, from their evolutionary origin to their current patterns. The main environmental drivers of mangrove ecology and evolution are also analyzed, with emphasis on plate tectonics and continental drift, climatic and sea-level shifts, and recent anthropogenic pressure. This is conducted using a time-continuum integrative approach aimed at circumventing the past–present dissociation and the associated methodological and conceptual drawbacks.

1.1 A Mangrove Story

This book deals with mangrove ecosystems from the Neotropics, specifically from the Caribbean region (Fig. 1.1). Mangroves are of that kind of unusual tropical ecosystem that is dominated by one or a few species. Before knowing the tropics, my

V. Rull, *Origin and Evolution of Caribbean Mangroves*, Ecological Studies 252,
https://doi.org/10.1007/978-3-031-57612-6_1

idea of a forest was that of a pine or an oak woodland, where the homogeneous structural basis is provided by the monotonous accumulation of trees from a single species across thousands of square kilometers. What was my surprise when, in the early 1980s, I stepped foot in a tropical jungle for the first time (Fig. 1.2). And this happened, neither more nor less, in one of the richest Neotropical cloud forests, those that grow on the coastal range of the southern Caribbean shores (Steyermark and Huber 1978; Huber 1986), where you can count more than 50 different tree

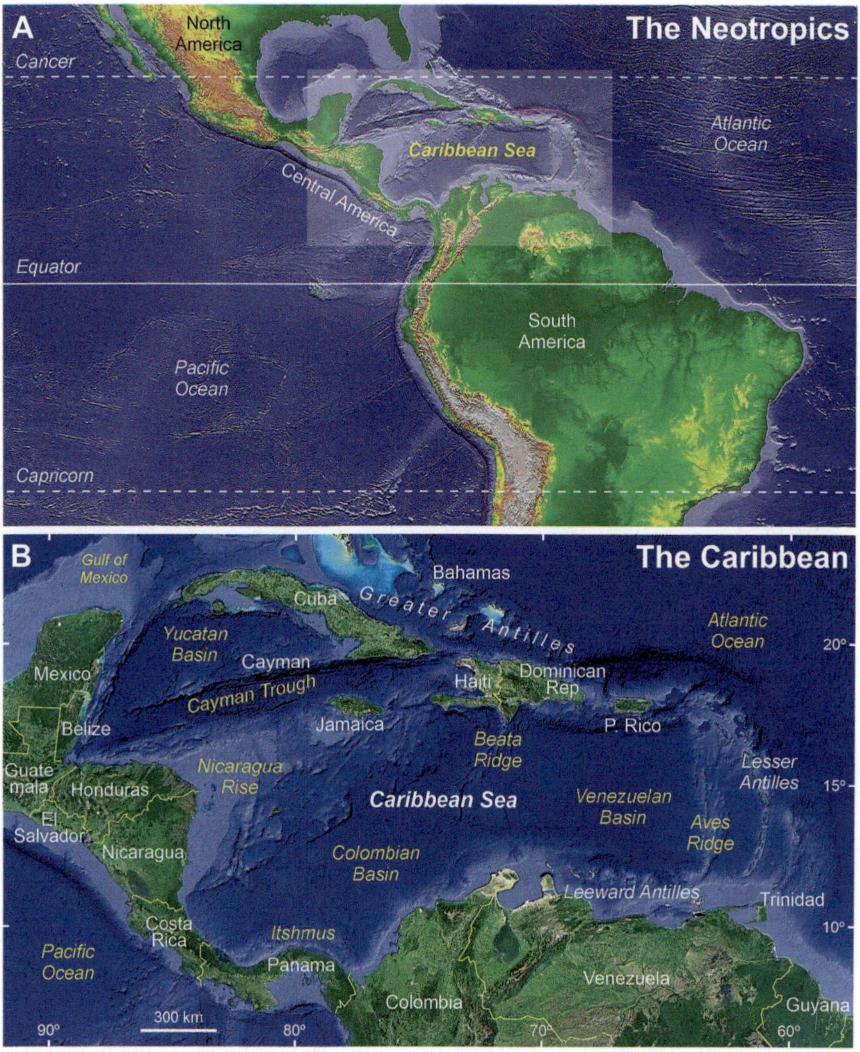

Fig. 1.1 The Caribbean region. (**a**) Map of the Neotropics with the Caribbean region marked by a transparent box. Base map from https://ngdc.noaa.gov/mgg/topo/globega2.html. (**b**) Google Earth image of the Caribbean region

Fig. 1.2 Examples of Neotropical rainforests from the Venezuelan Guiana region. (**a** and **b**) Gallery forests from the Gran Sabana uplands. (**c** and **d**) Montane forests from the Guiana highlands. Photos: V. Rull

species in the space of half-a-football pitch! I spent the next 20 years studying the long-term ecological dynamics of a variety of biomes ranging from the Caribbean coasts to the highest Andean summits, including the Orinoco lowlands and the Guiana uplands and highlands, growing under a wide range of environmental conditions, from desert to perhumid climates. This allowed me to verify by myself the classic idea of the tropics as a compendium of luxuriant communities with a huge variety of species, life forms, and ecological niches—along with colors, smells, sounds, and sensations—whose origin is still difficult to explain (Rull and Carnaval 2020). Another common characteristic of many Neotropical biomes is the high amount of endemic species, which may surpass 50% (Rull et al. 2019; Rull and Vegas-Vilarrúbia 2020) and whose explanation is equally elusive.

With time, I realized that the need for an explanation about purportedly unusual tropical biodiversity could be a human construction. Asking why the tropics are so diverse implicitly assumes that lower extratropical diversity is the norm and that the comparatively higher tropical richness needs an explanation. This is likely because, until recently, most ecological and evolutionary principles were developed in temperate regions and extrapolated to the tropics, where similar studies started later and were also conducted by researchers from temperate countries (Raby 2017). In the reverse situation, the question could have been why extratropical regions are so species poor (McGlone 1996), which is equally subjective. Actually, what needs an explanation is the so-called latitudinal diversity gradient (LDG), with maxima in the tropics and minima in the poles, which is perhaps the most general and conspicuous biodiversity pattern on Earth (Lomolino et al. 2010). The LDG was already noted by nineteenth-century scholars such as Johann Forster, Alexander von Humboldt, Alfred Wallace, or Charles Darwin, but a widely accepted explanation remains elusive (Pontarp et al. 2019), and it has been asked whether a general theory of tropical diversification actually exists or whether we should analyze the problem on a case-by-case, biome-by-biome, and/or region-by-region basis (Rull 2020). Even under this more general and objective perspective, however, the tropical variety of life is still amazing and challenging for any natural scientist, especially from extratropical regions.

Within this exuberant biological amalgamation, some peculiar ecosystems emerge that are dominated by one or a few species, such as the highland "superpáramos," which are dominated by a couple of high Andean endemic species of the compositae genus *Espeletia* (Luteyn 1999); the lowland/upland savannas from the Orinoco "llanos," dominated by a few grass species, usually from the genera *Trachypogon* and *Axonopus* (Huber 1987); or the lowland "morichales" (also "buritizales," "veredas," or "canangunchales," depending on the region), which are pure stands of *Mauritia*, a palm endemic to the Neotropics (Rull and Montoya 2014) (Fig. 1.3). Mangrove forests fall within this category but are even poorer, as they often lack understory and are less endemic, as the main dominant genera (*Rhizophora, Avicennia*) are of worldwide distribution (Tomlinson 2016). A common feature of all these communities is that they grow in extreme environments, where only plants with special adaptations are able to thrive. In the case of Andean highlands, the most limiting environmental constraints are extreme periglacial climates and soil instability, caused by freeze–thaw cycles (Monasterio 1980). The *Mauritia* palm stands are adapted to live in harsh conditions created by permanent flooding by extremely acidic waters (pH 4 or lower) and exceptionally nutrient-poor soils (Rull 1998a). Grass savannas are maintained and expanded by recurrent anthropogenic fires, to the detriment of rainforests (Rull et al. 2013, 2015). Mangroves live in extreme intertidal environments, with salinity as the main constraint but also influenced by permanent or intermittent flooding by marine waters (Lacerda et al. 2002).

Mangrove forests are dominated by a few mangrove-forming tree species—which in the Neotropics belong to the genera *Rhizophora* (Rhizophoraceae), *Avicennia* (Acanthaceae), and *Laguncularia* (Combretaceae) (Fig. 1.4)—characterized by their aerial roots (pneumatophores) and other morphological and physiological adaptations to live in intertidal environments submitted to daily flooding by seawaters (Tomlinson 2016). Besides these major elements, other minor but common components are *Pelliciera* (Tetrameristaceae), *Conocarpus* (Combretaceae), and the so-called mangrove fern *Acrostichum* (Pteridaceae). The floristics, biogeography, and ecology of mangroves are explained in more detail in the next chapter.

My scientific interest in mangroves was born in the late 1990s when I realized that the thick layers of sedimentary organic matter they accumulate contained long-term records of the ecological dynamics of these communities. This is possible thanks to the chronostratigraphic coherence of these deposits and their ability to preserve biological proxies—notably pollen, which is my favorite—due to their

Fig. 1.3 Examples of Neotropical ecosystems dominated by a single or a few species that define the main structural traits of the community. (**a**) High Andean páramo dominated by *Espeletia* (Colombia), publicly available at https://inaturalist.ca/observations/69173689; (**b**) upland morichal of *Mauritia* surrounded by grass savannas (Venezuela; photo V. Rull); (**c**) *Rhizophora* mangrove (Florida, USA), publicly available at https://inaturalist.ca/observations/46858758; (**d**) Detail of the aerial roots (pneumatophores) of *Rhizophora*, a fundamental structural mangrove feature (Florida, USA), publicly available at https://inaturalist.ca/observations/153735417

anoxic conditions (Ellison 2008). Using this type of evidence, I was able to reconstruct the Holocene sea-level rise on the southern Caribbean coasts (Rull 2000a; Rull and Vegas-Vilarrúbia 1999; Rull et al. 1999). By the time, I was working as an exploration biostratigrapher in a Venezuelan oil company and began to find significant amounts of mangrove pollen in rocks as old as Eocene (Rull 1998b, 1999, 2000b), which suggested me that it may be possible to assemble a continuous record of the evolution of Neotropical mangroves. Further exploration tasks on Oligocene to Pliocene rocks revealed that mangroves of these ages were more similar in composition to the present than to the Eocene ones (Rull 1997a, b, 2001, 2003). These observations led me to hypothesize that global environmental shifts that occurred during the Eocene/Oligocene transition (EOT) could have triggered an outstanding evolutionary change in Neotropical mangroves (Rull 1998c).

In the early 2000s, I came back to Europe and launched new research initiatives in other areas of the world, but a couple of years ago, I revisited the Caribbean mangrove evolution with the initial idea of gathering my former observations into a single coherent time-continuum sequence to follow the ecological and evolutionary history of mangroves, along with their biogeographical manifestations, from their

Fig. 1.4 The main Caribbean mangrove species. (**a**) *Rhizophora mangle* (red mangrove). (**b**) *Avicennia germinans* (black mangrove). (**c**) *Pelliciera rhizophorae* (tea mangrove). (**d**) *Laguncularia racemosa* (white mangrove). (**e**) *Acrostichum aureum* (mangrove fern). (**f**) *Conocarpus erectus* (buttonwood). Note the different morphologies of aerial roots. Reproduced from Rull (2022b)

evolutionary origin to the present. For this purpose, I assembled an extensive database encompassing not only my own records but also all Eocene ones to present original studies on mangroves published for the Caribbean region. The results of the analysis of this dataset have been published in a collection of papers organized chronologically (Rull 2022a, b, c, 2023a, b, c). This book synthesizes all this information in an attempt to build a consistent mangrove story under a time-continuum integrative perspective.

1.2 Importance of Mangroves

Mangroves are important for many reasons, but this is not essentially different from other ecosystems, and trying to magnify the relevance of ones over the others may be highly subjective and ephemeral and, therefore, useless, even nonsense. The reasons for considering mangroves as meaningful ecosystems may be intrinsic, functional, or practical. Intrinsic arguments are related to life itself and its evolution through time, whereas functional reasons are more linked to ecological dynamics and biogeochemical cycles, and practical explanations are mostly associated with human affairs such as survival and welfare. Only some important hints are provided here that will be developed in more depth in the next chapters.

All organisms and ecosystems have value in themselves by the mere fact of existing, with no need for instrumental or utilitarian reasons. This is called intrinsic value, and it is neither conferred nor revocable (Soulé 1985). From a scientific perspective, knowledge about these organisms and ecosystems, as part of knowledge advancement, also has intrinsic value, regardless of potential practical applications (Baker 1939; Rull 2016). Knowledge accumulation through history and its transmission from generation to generation are the basis for the maintenance of the identity and continuity of human culture; therefore, knowledge makes sense for itself (Rull 2014). There is no much more to say about this.

Ecologically, mangroves play a key role in the maintenance of continental and marine diversity and ecological dynamics, especially in coastal environments, as well as in global biogeochemical cycles, with an emphasis on the carbon cycle. These swamp forests form a distal continental fringe along tropical/subtropical coasts worldwide (approximately 25 °N to 25 °S), which marks the extent of normal (semidiurnal) tides (Fig. 1.5). The pneumatophores form a dense and complex network that efficiently traps continental and marine sediments (Fig. 1.6) and develops thick organic-rich deposits favoring coastal progradation under currently rising sea levels. Mangroves play a crucial role in safeguarding coastlines and various coastal ecosystems, such as corals, seagrasses, and salt marshes, by preventing erosion. They also create an intricate physical environment that promotes the diversification of habitats and niches. This, in turn, facilitates the thriving of numerous terrestrial and aquatic species, contributing to increased biodiversity and ecological complexity (Lugo and Snedaker 1974; Saenger 2002; Nagelkerken et al. 2008) (Fig. 1.7). A particular feature of mangroves is that they provide nursery habitats among their submerged roots, where food availability is maximized and predation incidence is minimized, which favors the survival of juvenile stages of many marine species (Laegdsgaard and Johnson 2001) (Fig. 1.7).

From a pragmatic human standpoint, mangroves offer not just ecological and cultural services such as fisheries, agriculture, aquaculture, timber, fuel, aesthetics, and ecotourism (review in Afonso et al. 2021), but they also play a role in mitigating global warming. In fact, these ecosystems, alongside seagrasses and salt marshes, stand out as crucial blue-carbon ecosystems, which effectively capture and sequester carbon, thereby extracting it from the global carbon cycle. This function makes them proficient carbon sinks, aiding in the reduction of atmospheric CO_2 levels (Nellemann et al. 2009; Mcleod et al. 2011; Fest et al. 2022). Terrestrial forests store carbon in two main forms: as aboveground living biomass and as belowground organic accumulations. Contrary to terrestrial forests, which accumulate carbon mainly in the living compartment, most carbon sequestered by mangroves is buried in their organic-rich sediments. It has been estimated that mangroves immobilize >10 times more carbon in their sediments than tropical, temperate, and boreal forests (Mcleod et al. 2011) (Fig. 1.8).

Nevertheless, mangroves face significant threats as one of the most endangered ecosystems globally (Worthington et al. 2020). Recent assessments indicate that the worldwide mangrove coverage has diminished by 3.4% within a span of less than 25 years (1996–2020) due to both natural factors and human-induced deforestation (Bunting et al. 2022). Should these depletion rates persist, there is a great risk of substantial decline in mangrove ecosystems over the course of this century, posing a serious threat to their long-term existence (Duke et al. 2017), which would imply

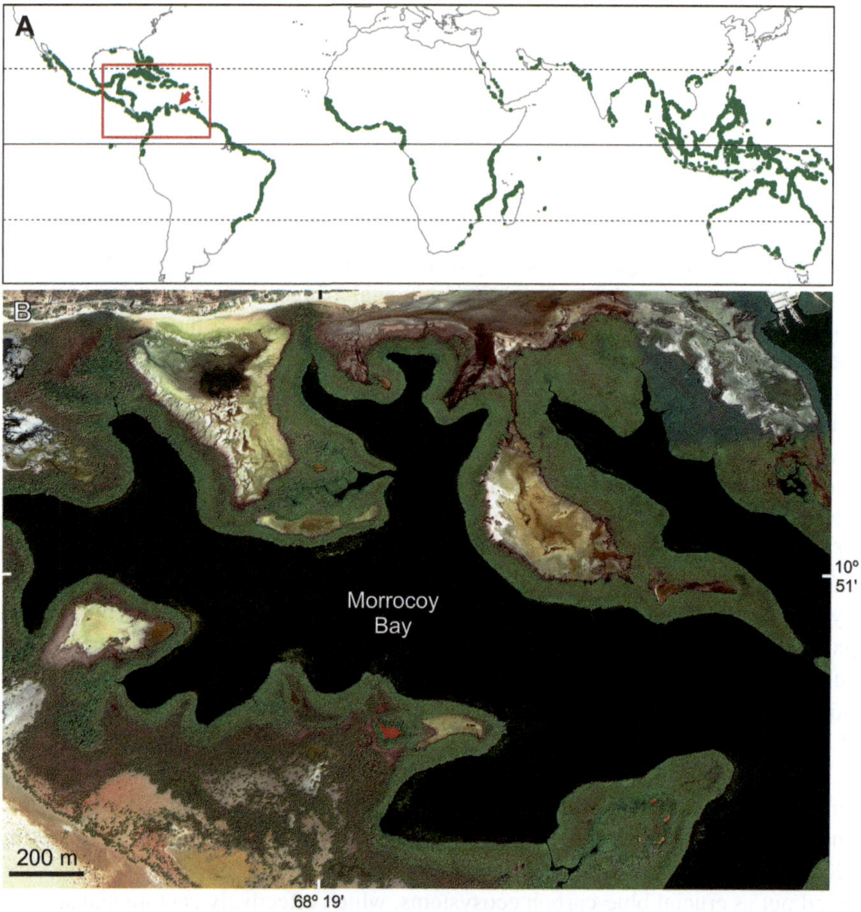

Fig. 1.5 The mangrove fringe. (**a**) World distribution of mangroves (green) indicating the Caribbean region (red box) and the location of Morrocoy National Park (red arrow). Base map from Spalding et al. (2010). (**b**) Coastal mangrove fringe (green) in Morrocoy National Park (Venezuela), situated in the southern Caribbean coasts, when I had my first contact with Caribbean mangroves, in 1981. Base image: Google Earth

a significant loss in the immobilized carbon stock (Fig. 1.9) and hence in the global warming mitigation capacity. This has fostered the launching of numerous worldwide initiatives for mangrove conservation and restoration, which need sound ecological and evolutionary knowledge (Makowski and Finkl 2018; Lacerda et al. 2019; Lester et al. 2020; Mishra and Farooq 2022).

In addition to contemporary ecological research, the extended historical records spanning centuries and millennia prove particularly valuable. These records offer direct observational evidence of how mangrove ecosystems have responded to various environmental factors such as climate, sea level changes, and human activities. This firsthand data aids in identifying key threats and establishing response

Fig. 1.6 Suspended sediments and sediment accumulation within mangrove roots. (**a**) *Rhizophora* mangrove from Costa Rica, publicly available at https://inaturalist.ca/observations/131132414; (**b**) *Avicennia* mangrove from Florida (USA), publicly available at https://inaturalist.ca/observations/41407485

thresholds, offering valuable insights for effective mangrove management. Additionally, evolutionary studies contribute significant evidence by highlighting the adaptive potential of mangrove species and their ability to undergo genetic changes in response to environmental shifts (Rull 2023c).

1.3 The Time-Continuum Integrative Approach

In addition to the abovementioned myth of tropical biodiversity, another human construction is the profound abyss existing between past and present, which is especially manifest in ecological research. I discussed this topic in depth more than a

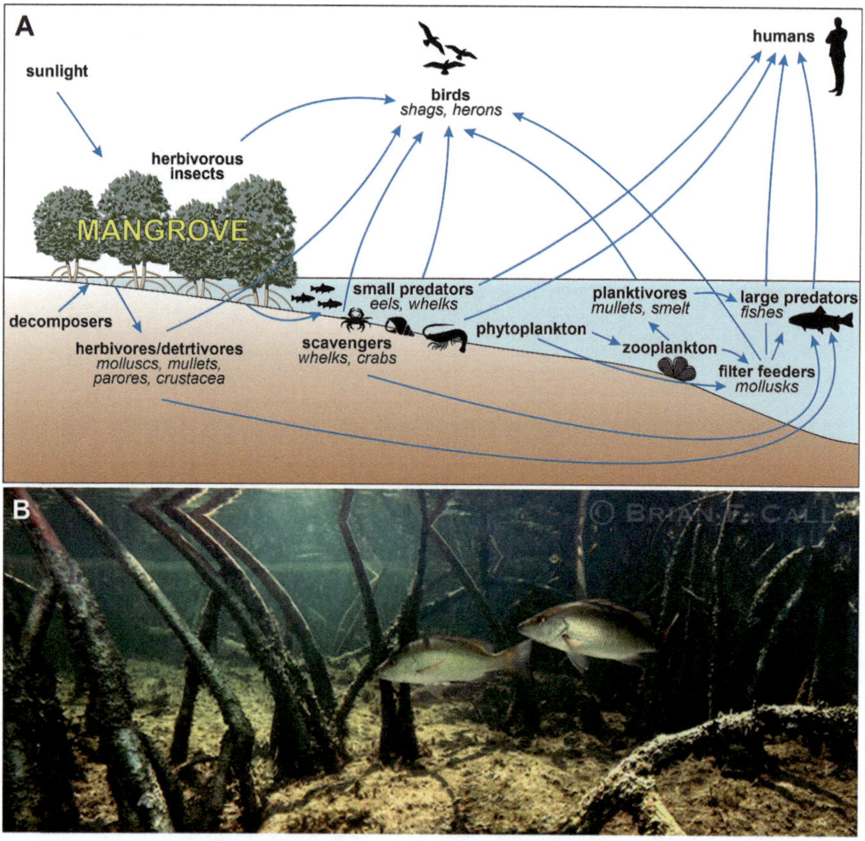

Fig. 1.7 The mangrove ecosystem. (**a**) Schematic diagram of a Neotropical mangrove food web; (**b**) Mangrove snappers (*Lutjanus griseus*) swimming among submerged mangrove roots in the Florida Everglades (USA), publicly available at https://commons.wikimedia.org/wiki/File:Mangrove_snapper,_Lutjanus_griseus.jpg

decade ago (Rull 2010), and only the most important clues are disclosed here. The basic tenet is that there is one single biosphere that has remained since the origin of life on Earth; there is no a biosphere of the past and another of the present. The old Cuvier's pseudoscientific theories of recurrent global catastrophes and creation acts (Rudwick 1997) have already been overthrown, and now, we know that the extinction crises that occurred during the history of Earth (MacLeod 2015) have led to deep reorganizations but always within the same biosphere. This biospheric time continuum may be decomposed into three main components: the ecological continuum, the evolutionary continuum, and the biogeographical continuum.

First, it is well-known that life is not possible without ecological interactions. In the words of the eighteenth-century Scottish anatomist John Hunter, "Nothing in nature stands alone" (Bryant 1893). Even the simplest life form should be part of a matter/energy cycle to persist. Good examples are the stromatolites, which are

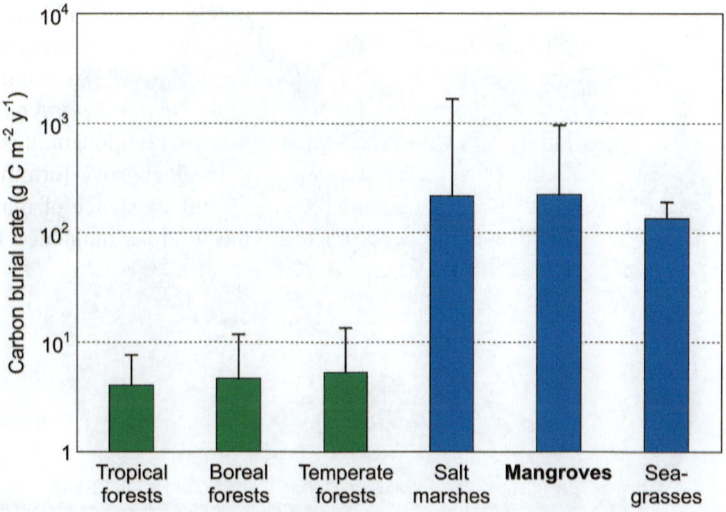

Fig. 1.8 Mean rates and standard errors of carbon sequestration in soils and sediments from different terrestrial (green bars) and coastal marine (blue bars) ecosystems. Note the logarithmic scale of the y-axis. Redrawn from Mcleod et al. (2011)

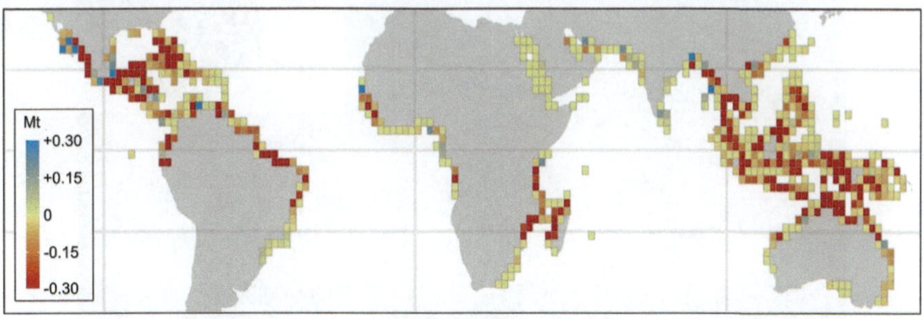

Fig. 1.9 Global change in mangrove carbon stock due to deforestation between 1996 and 2016. Colors represent the carbon balance between forestation and deforestation in million tons (Mt). Modified from Richards et al. (2020)

among the most ancient known life forms (Tewari and Seckbach 2011) and are formed by two microbial layers, one photosynthetic and the other heterotrophic, conforming a very simple production-consumption-decomposition cycle supported by energy (light), water, and nutrient inputs from the environment (Fig. 1.10). Second, biological evolution is also a continuous process that has not stopped since the origin of life and has been responsible for the turnover of the ecological actors over time (Hutchinson 1965). For example, the terrestrial Jurassic/Cretaceous world was dominated by large reptiles, whereas the Cenozoic world has been dominated by large mammals, but the basic ecological rules are the same. Indeed, as formerly proposed by Lyell (1850), natural laws are invariable in space and time, a rule that is now known as the principle of uniformitarianism (Romano 2015). Third, the

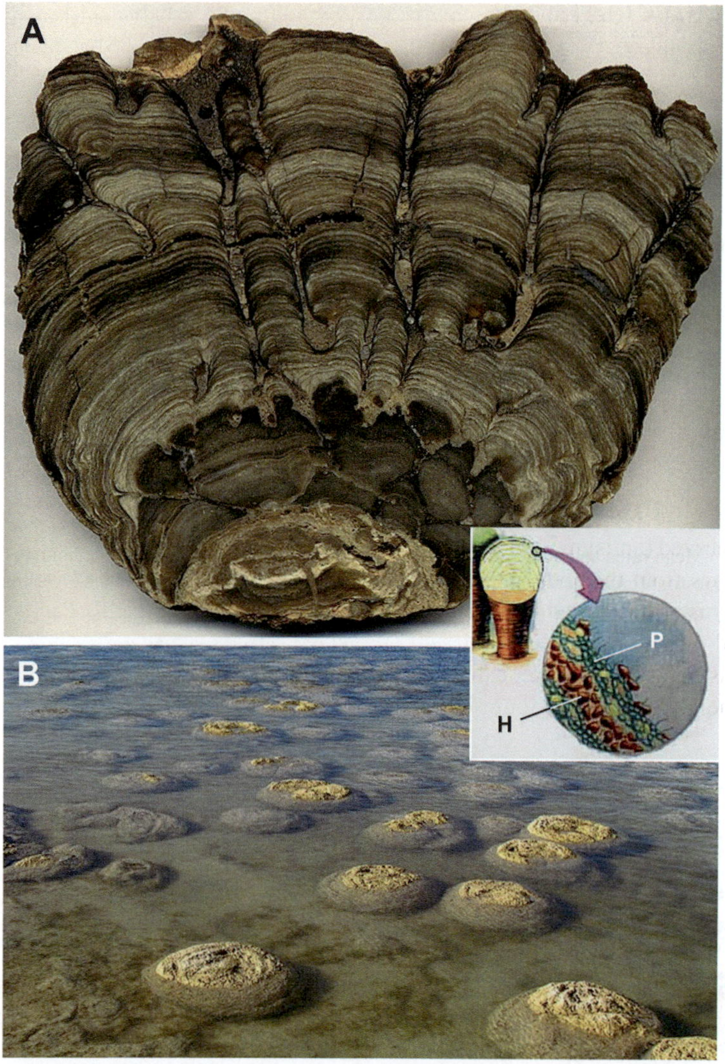

Fig. 1.10 Fossil and living stromatolites. (**a**) Fossil stromatolite from the Eocene of Wyoming (USA) showing the typical bilayered structure. Publicly available at https://commons.wikimedia.org/wiki/File:Stromatolite_Fossil_from_Wyoming.jpg; (**b**) living stromatolites from Australia. Publicly available at https://commons.wikimedia.org/wiki/File:Stromatolithes1.jpeg. The diagram shows an idealized cross section of the external layers (P, photosynthetic layer; H, heterotrophic layer)

interactions of the biosphere with other compartments of the Earth system (atmosphere, hydrosphere, and lithosphere) have been not only important evolutionary engines and ecological modulators but also the primary cause for the continuous shifts in the distribution of living beings across the planet or biogeographical reorganizations. Therefore, the biospheric continuum relies on the existence of ecological, evolutionary, and biogeographical continua that have followed the same rules since the establishment of life on Earth. In this context, the separation between past and present does not make sense, at all.

Past-present dissociation is more a by-product of human limitations, firmly rooted in the nature of evidence and the ensuing methodological constraints, which largely determine the scientific disciplines participating in each study (Rull 2010). For example, fossil stromatolites are usually studied by paleontologists using geological methods because they are actually rocks, whereas extant stromatolites, as living beings, are studied primarily by ecologists using biological methods. This methodological disparity, which is bounded to the nature of the evidence, leads to radical differences, not only in the way that stromatolites are approached but also in the ultimate research objectives. However, the study target is the same, the stromatolites, and the sum of all approaches is needed to understand them within an integrative ecological/evolutionary/biogeographical, or biospheric, continuum framework. In recent centuries, knowledge accumulation has progressed in a highly specialized fashion, and ancient naturalists who were skilled in a wide range of disciplines have already gone. This is frequently attributed to the comparatively huge amount of extremely specialized information available today, which is impossible to capture and process by a single person. In this scenario, the most useful strategy is to assemble multidisciplinary teams to analyze the evidence from a holistic perspective. This procedure works within the past and present research communities separately; however, gathering specialists on past and present biospheric affairs to work together is incredibly difficult. To put it very simply, many paleontologists/paleoecologists consider that the time frame of modern ecological studies is too short to be important, whereas many modern ecologists consider the past as an interesting, yet ecologically irrelevant, museum-like curiosity. Overcoming these narrow-minded conceptions is essential for the progress of integrative knowledge and for understanding our biosphere, which is the ultimate scientific goal (Rull 2010).

This book is a trial to understand the Caribbean mangroves from a time- and discipline-transgressive continuum biospheric perspective, to encourage the development of similar attempts in other mangrove-bearing regions worldwide, as well as in other biomes and ecosystems. In this attempt, I have tried to take advantage of my professional experience that combines ecological, paleoecological, and evolutionary expertise, but I am fully aware that a single person will hardly be able to fully achieve such goal. Therefore, I will be happy enough if this book contributes to setting the basis for a better understanding of Caribbean mangroves and generates interest in the time-continuum integrative approach and its potential to properly understand natural systems. Terms such as "geological time," "deep time," "Q-time," "real time," "ecological time," and others similar (Jackson 2001) that accentuate the unreal anthropocentric disparity between past and present are avoided here, to emphasize the continuous nature of time and its ecological and biogeographical manifestations during the evolution of life on Earth (Rull 2012).

1.4 Temporal Scope

The temporal framework of this book is the Cenozoic era (66 Ma to present), with incursions into the Late Cretaceous (Maastrichtian; 72–66 Ma), when discussing the origin of Caribbean and Neotropical mangroves (Fig. 1.11). Ages are expressed in million years before present (Ma), except for the Quaternary (the last 2.6 Ma), especially the Holocene (the last 11,700 years), where ages are typically given in years (yr) or kiloyears (kyr) before present (BP). The main paleogeographic, paleoclimatic, and paleoeustatic changes that occurred during the Cenozoic are depicted in Figs. 1.12 and 1.13, which show the major trends at a global scale to provide a general framework of the main environmental drivers. The particular geographical, climatic, and sea-level shifts of the Caribbean region, along with their influence on mangrove ecology, evolution, and biogeography, will be discussed in the corresponding chapters at higher spatiotemporal resolution. However, some major features have been particularly relevant for mangrove evolution in the Neotropics and deserve special mention.

For example, the passage from the Cretaceous to the Paleogene was characterized by the progressive closure of the pantropical Tethys Sea by the African continental barrier (Fig. 1.12), which would have initiated the split of eastern and western mangrove floras that still characterize the present global biogeography of this biome (McCoy and Heck 1976; Ellison et al. 1999). Another critical phase would have been the Eocene/Oligocene transition (EOT), characterized by global cooling and a significant sea-level drop (Fig. 1.13), which have been linked to the consolidation of modern mangrove biogeography (Plaziat et al. 2001; Srivastava and Prasad 2018) and major evolutionary turnover in New World mangroves (Rull 1998c). Before the EOT, the Earth was in a Hothouse/Warmhouse state, and the poles were devoid of ice, but the Oligocene was the beginning of a Coolhouse/Icehouse Earth initiated with the glaciation of Antarctica (Figs. 1.11 and 1.12). The Northern Hemisphere, including the North Pole, was covered by ice only in the last couple of million years, during the Pleistocene. This long-term Neogene and Quaternary cooling, spiked by frequent and intense sea-level fluctuations, was paralleled by a progressive diversification of Neotropical mangroves (Graham 1995), although eventual causal relationships between these environmental and biotic shifts remain unknown.

How Pleistocene glaciations affected Caribbean mangroves is also largely unknown due to the lack of records long enough to fully encompass this geological period. It is commonly assumed that extant mangrove species were already present at the onset of the Quaternary and that the ensuing climatic and sea-level glacial-interglacial fluctuations fostered spatial reorganization, rather than diversification or biodiversity depletion (Graham 1995). However, a comprehensive synthetic analysis to verify this hypothesis is still unavailable. Human pressure on Neotropical mangroves began later than in other tropical areas due to the later settlement of the American continent by human societies, which did not occur until the Late Pleistocene, at the end of the last glacial cycle (O'Brien 2019). However, the last millennia have witnessed an intensification of anthropogenic disturbance of

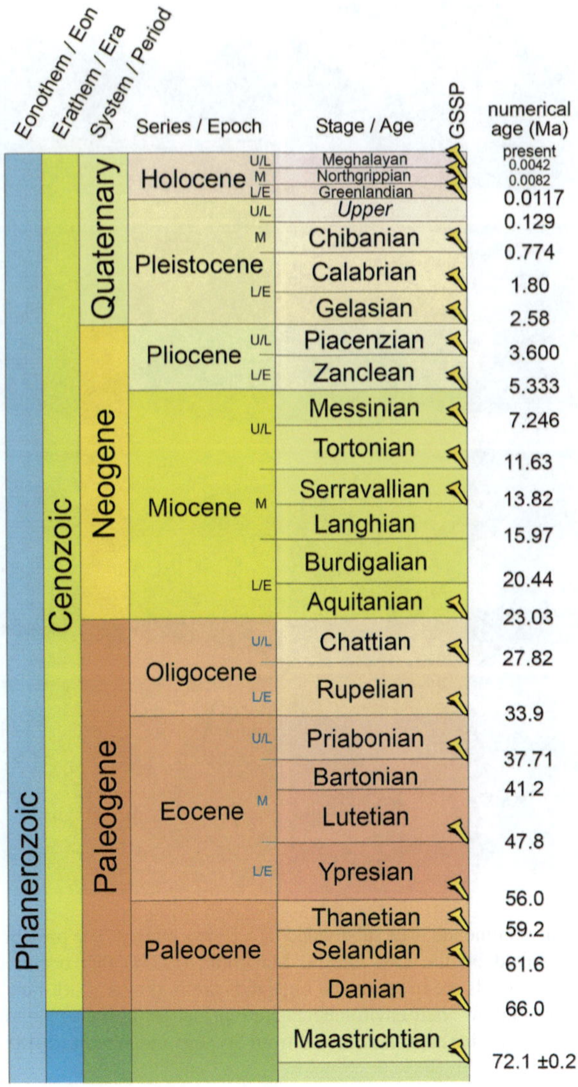

Fig. 1.11 Updated (v2022/10) version of the International Chronostratigraphic Chart (ICS) corresponding to the Cenozoic era. Eocene and Oligocene L/E (Lower/Early), M (Middle), and U/L (Upper/Late) subdivisions (in blue) have been added as informal but commonly used units. GSSP, Global Boundary Stratotype Sections and Points. Based on Cohen et al. (2013)

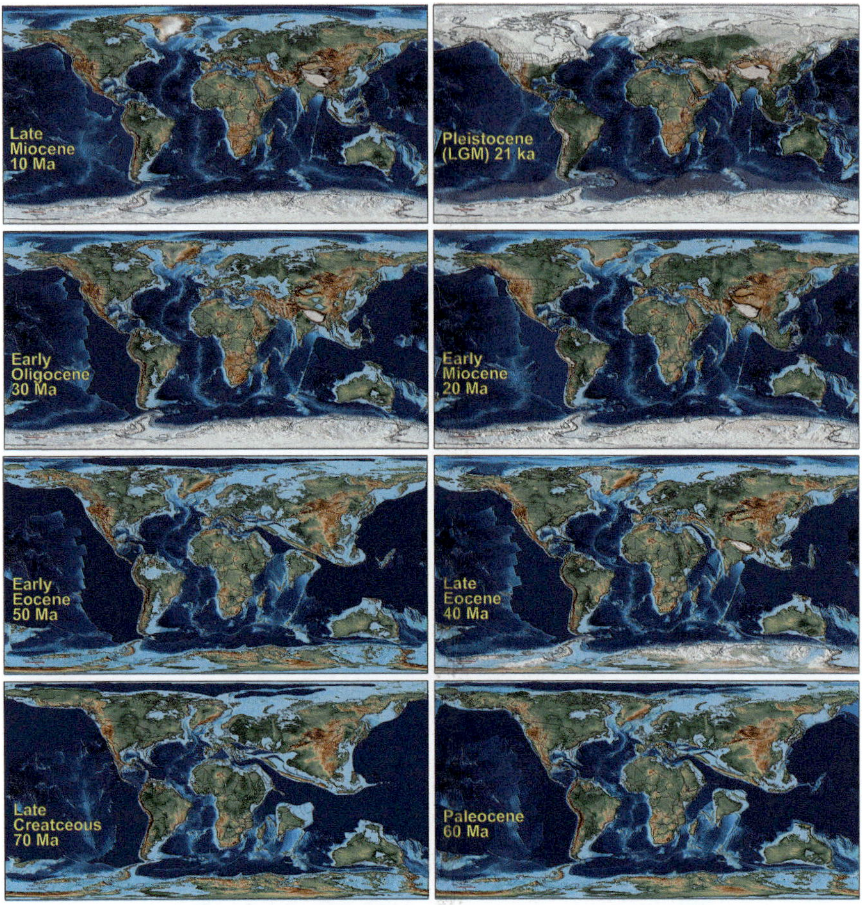

Fig. 1.12 Cenozoic continental drift subdivided into 10-my phases. The present continental configuration is represented by the Last Glacial Maximum (LGM) that occurred approximately 21,000 years ago. Lowlands are in green and highlands are in brown. Dark-blue represents deep-marine settings, and shallow shelf environments are represented in light-blue tones. White/light-gray tones represent major ice caps. Assembled from Scotese and Wright (2018)

Caribbean mangroves similar to that observed worldwide (López-Angarita et al. 2016).

1.5 Mangrove or Mangal?

Before getting fully into the study of Caribbean mangroves, there is a terminological issue that should be clarified, as is the mangal/mangrove controversy. Historically, the term "mangrove" has been used in either a collective or individual sense to refer to this particular biome/vegetation type, its communities, or its constituent plant

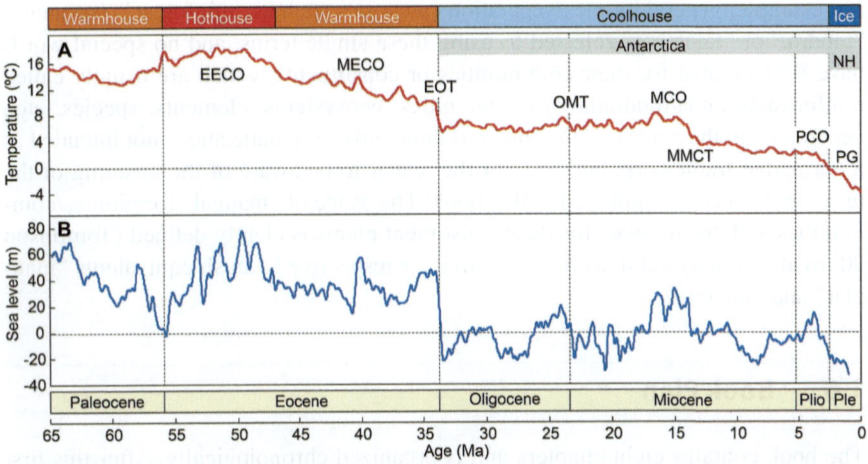

Fig. 1.13 Cenozoic shifts in temperature (**a**) and sea level (**b**), expressed as anomalies with respect to the present. The glaciation of the south (Antarctica) and north (NH) poles is indicated by gray bands. Ice, Icehouse; Plio, Pliocene; Ple, Pleistocene; EECO, Early Eocene Climate Optimum; MECO, Middle Eocene climate optimum; EOT, Eocene/Oligocene transition; OMT, Oligocene/Miocene transition; MCO, Miocene climate optimum; MMCT, Middle Miocene climate transition; PCO, Pliocene climate optimum; PG, Pleistocene glaciation. Temperature reconstruction after Westerhold et al. (2020) and eustatic curve after Miller et al. (2020)

species (Tomlinson 2016). To differentiate these concepts, MacNae (1968) introduced the term "mangal" for the biome/community leaving "mangrove" for its constituent species. In this framework, for example, this book would be about the Caribbean "mangals," rather than "mangroves," whereas the *Rhizophora*, *Avicennia*, and *Laguncularia* species would be "mangroves." This terminology is consistent with popular usage in Spanish, the most widespread language in the Caribbean region, where a mangrove community is called "manglar," whereas a mangrove tree is known as a "mangle." This has led to the popular names "mangle rojo/red mangrove" for *Rhizophora*, "mangle negro/black mangrove" for *Avicennia*, "mangle blanco/white mangrove" for *Laguncularia*, and "mangle piñuelo/tea mangrove" for *Pelliciera*. However, the term "mangal" has not been firmly established in the English scientific literature, and the term "mangrove" is still generally used in both collective and individual senses. In addition, the terms "mangrove" and "mangal" are sometimes incorrectly utilized as synonyms, which creates confusion. The same is true for the Spanish terms "manglar" and "mangle."

In this book, the term "mangal" is avoided, and "mangrove" alone is used to refer to this particular vegetation type or biome, rather than the community or its constituents, which are named using composite terms, such as "mangrove forest," "mangrove ecosystem," "mangrove community," "mangrove tree," "mangrove element," "mangrove associate," "mangrove species," and others similar. This is a usual procedure in biogeography, in general, and in vegetation science, in particular, especially for major worldwide biomes and vegetation types such as mangroves.

For example, general biomes/vegetation categories such as "rainforest," "savanna," "tundra," or "taiga" are referred to using these single terms, and no special words have been created for their communities or constituents, which are usually called rainforest/savanna/tundra/taiga communities, ecosystems, elements, species, etc., depending on the context. This choice is eminently pragmatic and is not intended as a normative framework, but only for the reader to be aware of the meaning of the terms and concepts employed in this book. The usage of "mangal" for biomes/communities and "mangrove" for their constituent plants is clearly defined (Tomlinson 2016) and equally valid, as occurs with their respective Spanish equivalents "manglar" and "mangle."

1.6 Book Plan

The book contains eight chapters and is organized chronologically. After this first introductory account, the second chapter summarizes the present-day state of the Caribbean mangroves as plant communities. The first part of the chapter briefly introduces the main environmental features enveloping these Neotropical coastal forests, and the second part succinctly describes their extent, floristic composition, biogeography, and main ecological traits. The third chapter addresses the origin of Caribbean mangroves, focused on the debate about when and how these ecosystems emerged. This topic is discussed under a community approach using quantitative data, aimed at unraveling the origin of mangroves as ecosystems, rather than identifying individual mangrove elements in the fossil record, which is the most usual procedure. Using this ecosystem approach, a picture that strongly contrasts, in terms of both tempo (Eocene) and mode (de novo origin), with the traditional view (Early Cenozoic origin by regional differentiation of a Late-Cretaceous pantropical mangrove belt) is obtained. The fourth chapter deals with the most intense evolutionary turnover undergone by the Caribbean mangroves, consisting of the replacement of the original Eocene *Pelliciera*-dominated mangrove communities by the Oligocene ones dominated by *Rhizophora* (the present-day dominant), a newly emerged species that was absent in the primeval Eocene mangroves. This evolutionary revolution coincided with the significant EOT global biotic turnover, characterized by large-scale extinctions and increased radiations, coeval with the shift from a greenhouse to an icehouse Earth state. Chapter 5 is about the long Neogene evolutionary stasis recorded since the major EOT evolutionary turnover, which was followed by a relatively recent (Pliocene) diversification burst that shaped the extant Caribbean mangrove communities. The influence of the Pleistocene environmental (climatic and eustatic) shifts and the impact of Holocene human activities, along with the corresponding climatic-anthropogenic feedbacks and synergies, are discussed in Chap. 6. Chapter 7 is an evolutionary synthesis centered on the *Pelliciera–Rhizophora* interactions that define a taxon cycle for the first and suggests a potential explanation for the EOT turnover from a long-term time-continuum ecological perspective. Chapter 8 is a summary of the main conclusions obtained, along with a discussion on the potential applications that paleoecological and evolutionary studies may have for mangrove conservation and restoration.

References

Afonso F, Félix PM, Chainho P et al (2021) Assessing ecosystem services in mangroves: insights from São Tomé Island (Central Africa). Front Environ Sci 9:501673

Baker JR (1939) A counterblast to Bernalism. New Statesman Nat 18:174–175

Bryant T (1893) Hunterian oration. Lancet 1:342

Bunting P, Rosenqvist A, Hilarides L et al (2022) Global Mangrove Watch: updated 2010 mangrove forest extent (v2.5). Remote Sens 10:1669

Cohen KM, Finney SC, Gibbard PL et al (2013) The ICS international chronostratigraphic chart. Episodes 36:199–204. (version 2020/10; regularly updated at www.stratigraphy.org/ICSChart/)

Duke NC, Meyneccke J-O, Dittman S et al (2017) A world without mangroves? Science 317:41b–42b

Ellison JC (2008) Long-term retrospection on mangrove development using sediment cores and pollen analysis: a review. Aquat Bot 89:93–104

Ellison AM, Farnsworth EJ, Merkt RE (1999) Origins of mangrove ecosystems and the marine biodiversity anomaly. Glob Ecol Biogeogr 8:95–115

Fest BJ, Swearer SE, Arndy SK (2022) A review of sediment carbon sampling methods in mangroves and their broader impacts on stock estimates for blue carbon ecosystems. Sci Total Environ 816:151618

Graham A (1995) Diversification of Gulf/Caribbean mangrove communities through Cenozoic time. Biotropica 27:20–27

Huber O (1986) La Selva Nublada de Rancho Grande, Parque Natural Henri Pittier. Editorial Arte, Caracas

Huber O (1987) Neotropical savannas: their flora and vegetation. Trends Ecol Evol 2:67–71

Hutchinson GE (1965) The ecological theater and the evolutionary play. Yale University Press, New Haven

Jackson JB (2001) Integrating ecological dynamics across time-scales, realtime, Q-time and deep time. PALAIOS 16:1–2

Lacerda LD, Conde JE, Kjerve B et al (2002) American mangroves. In: Lacerda LD (ed) Mangrove ecosystems. Springer, Berlin, pp 1–62

Lacerda LD, Borges R, Ferreira AC (2019) Neotropical mangroves: conservation and sustainable use in a scenario of global climatic change. Aquat Conserv 29:1347–1364

Laegdsgaard P, Johnson C (2001) Why do juvenile fish utilize mangrove habitats. J Exp Mar Biol Ecol 257:229–253

Lester SE, Dubel AK, Hernán G et al (2020) Spatial planning principles for marine cosystem restoration. Front Mar Sci 7:328

Lomolino MV, Riddle BR, Brown JH et al (2010) Biogeography. Sinauer, Sunderland

López-Angarita J, Roberts CM, Tilley A et al (2016) Mangroves and people: lessons from a history of use and abuse in four Latin American countries. Forest Ecol Manag 368:1151–1162

Lugo AE, Snedaker SC (1974) The ecology of mangroves. Ann Rev Ecol Syst 5:39–64

Luteyn JL (1999) Páramos: a checklist of plant diversity, geographical distribution, and botanical literature. The New York Botanical Garden Press, New York

Lyell C (1850) Principles of geology. John Murray, London

MacLeod N (2015) The great extinctions: what causes them and how they shape life. Firefly, Richmond Hill

MacNae W (1968) A general account of the fauna and flora of mangrove swamps and forests in the Indonesian West Pacific. Adv Mar Biol 6:73–270

Makowski C, Finkl CW (2018) Threats to Mangrove forests. Hazards, vulnerability, and management. Springer Nature, Cham

McCoy ED, Heck KL (1976) Biogeography of corals, seagrasses, and mangroves: an alternative to the center of origin concept. Syst Zool 25:201–210

McGlone MS (1996) When history matters: scale, time, climate and tree diversity. Glob Ecol Biogeogr 5:309–314

Mcleod E, Chmura GL, Bouillon S et al (2011) A blueprint for blue carbon: toward an improved understanding of the role of vegetated coastal habitats in sequestering CO_2. Front Ecol Environ 9:552–560
Miller KG, Browning JV, Schmelz WJ et al (2020) Cenozoic sea-level and cryospheric evolution from deep-sea geochemical and continental margin records. Sci Adv 6:aaz1346
Mishra AK, Farooq SH (2022) Lack of ecological data hinders management of ecologically important saltmarsh ecosystems: a case study of saltmarsh plant *Porterasia coarctata* (Roxb.). J Environ Manag 321:115957
Monasterio M (1980) Estudios Ecológicos en los Páramos Andinos. Univ Los Andes, Mérida
Nagelkerken I, Blaver SJN, Bouillon S et al (2008) The habitat function of mangroves for terrestrial and marine fauna: a review. Aquat Bot 89:155–185
Nellemann C, Corcoran E, Duarte C et al (2009) Blue carbon. A rapid response assessment. UNEP, GRID-Arendal
O'Brien MJ (2019) Setting the stage: the Late Pleistocene colonization of North America. Quaternary 2:1
Plaziat J-C, Cavagnetto C, Koeniguer J-C et al (2001) History and biogeography of the mangrove ecosystem, based on a critical reassessment of the paleontological record. Wetl Ecol Manag 9:161–179
Pontarp M, Bnnefeld L, Cabral JS et al (2019) The latitudinal diversity gradient: novel understanding through mechanistic eco-evolutionary models. Trends Ecol Evol 34:211–223
Raby M (2017) The colonial origin of tropical field stations. Am Sci 105:216–223
Richards DR, Thompson BS, Wijedasa L (2020) Quatifying net loss of mangrove carbon stocks from 20 years of land cover change. Nat Commun 11:4260
Romano M (2015) Reviewing the term uniformitarianism in modern Earth Sciences. Earth-Sci Rev 148:65–76
Rudwick MJS (1997) Geoges cuvier, fossil bones, and geological catastrophes. University Chicago Press, Chicago
Rull V (1997a) Oligo-Miocene palynology of the Rio Chama sequence (western Venezuela), with comments on fossil algae as paleoenvironemntal indicators. Palynology 21:213–229
Rull V (1997b) Sequence analysis of western Venezuelan Cretaceous to Eocene sediments using palynology: chrono-paleoenvironmental and paleovegetational approaches. Palyology 21:79–90
Rull V (1998a) Biogeographical and evolutionary considerations on *Mauritia* (Arecaceae), based on palynological evidence. Rev Palaeobot Palynol 100:109–122
Rull V (1998b) Middle Eocene mangroves and vegetation changes in the Maracaibo basin, Venezuela. PALAIOS 13:287–296
Rull V (1998c) Evolución de los manglares neotropicales: la crisis del Eoceno. Interciencia 23:355–362
Rull V (1999) Palaeofloristic and palaeovegetational changes across the Paleocene/Eocene boundary in northern South America. Rev Palaeobot Palynol 107:83–95
Rull V (2000a) Holocene sea level rise in Venezuela: a preliminary curve. Bol Soc Ven Geol 25:32–36
Rull V (2000b) Ecostratigraphic study of Paleocene and Early Eocene palynological cyclicity in northern South America. PALAIOS 15:14–14
Rull V (2001) A quantitative palynological record from the Early Miocene of western Venezuela, with emphasis on mangroves. Palynology 25:109–126
Rull V (2003) Contribution of quantitative ecological methods to the interpretation of stratigraphically homogeneous pre-Quaternary sequences: a palynological example from the Oligocene of Venezuela. Palynology 27:75–98
Rull V (2010) Ecology and palaeoecology: two approaches, one objective. Open Ecol J 3:1–5
Rull V (2012) Time, evolution and physical reductionism. EMBO Rep 13:181–185
Rull V (2014) The most important application of science. EMBO Rep 15:919–922
Rull V (2016) Free science under threat EMBO rep 17:131–135

Rull V (2020) Neotropical diversification: historical overview and conceptual insights. In: Rull V, Carnaval A (eds) Neotropical diversification: patterns and porcesses. Springer Nature, Cham, pp 13–49
Rull V (2022a) The Caribbean mangroves: an Eocene innovation with no Cretaceous precursors. Earth-Sci Rev 231:104070
Rull V (2022b) Responses of Caribbean mangroves to Quaternary climatic, eustatic and anthropogenic drivers of ecological change: a review. Plan Theory 11:3502
Rull V (2022c) Eocene/Oligocene global disruption and the revolution of Caribbean mangroves. Persp Plant Ecol Evol Syst 59:125733
Rull V (2023a) Taxon cycles in Neotropical mangroves. Plan Theory 12:244
Rull V (2023b) The Neogene-Quaternary diversification trend in the shaping of modern Caribbean mangroves. Quat Sci Rev 300:107920
Rull V (2023c) Rise and fall of Caribbean mangroves. Sci Total Environ 885:163851
Rull V, Carnaval A (2020) Neotropical diversification: patterns and processes. Springer Nature, Cham
Rull V, Montoya E (2014) *Mauritia flexuosa* palm swamp communities: natural or human-made? A palynological study of the Gran Sabana region (northern South America) within a Neotropical context. Quat Sci Rev 99:17–33
Rull V, Vegas-Vilarrúbia T (1999) Surface palynology of a small coastal basin from Venezuela and potential paleoecological applications. Micropaleontology 45:365–393
Rull V, Vegas-Vilarrúbia T (2020) The Pantepui "Lost World": towards a biogeographical, ecological and evolutiorany synthesis of a pristine Neotropical sky-island archipelago. In: Rull V, Carnaval A (eds) Neotropical diversification: patterns and porcesses. Springer Nature, Cham, pp 369–413
Rull V, Vegas-Vilarrúbia T, Espinoza N (1999) Palynological record of an Early-Mid Holocene mangrove in eastern Venezuela. Implications for sea-level rise and disturbance history. J Coast Res 15:496–504
Rull V, Montoya E, Nogué S et al (2013) Ecological palaeoecology in the Neotropical Gran Sabana region: long-term records of vegetation dynamics as a basis for hypothesis testing. Persp Plant Ecol Evol Syst 15:338–359
Rull V, Montoya E, Vegas-Vilarrúbia T et al (2015) New insights on palaeofires and savannisation in northern South America. Quat Sci Rev 122:158–165
Rull V, Vegas-Vilarrúbia T, Huber O et al (2019) Biodiversity of Pantepui. The pristine "lost world" of the neotropical Guiana highlands. Academic Press/Elsevier, London
Saenger P (2002) Mangrove ecology, silviculture and conservation. Kluwer, Dordrecht
Scotese CR, Wright N (2018) PALEOMAP Paleodigfital Elevation Models (PaleoDEMS) for the Phanerozoic. Zenodo, https://doi.org/10.5281/zenodo.5348492 (free downloads at Earthbyte; https://www.earthbyte.org/paleodem-resource-scotese-and-wright-2018/)
Soulé ME (1985) What is conservation biology? Bioscience 35:727–734
Spalding M, Kainuma M, Collins L (2010) World Atlas of Mangroves. Routledge, London
Srivastava J, Prasad V (2018) Evolution and paleobiogeography of mangroves. Mar Ecol 40:e12571
Steyermark JA, Huber O (1978) Flora del Avila. Soc Ven Cien Nat, Caracas
Tewari V, Seckbach J (2011) Stromatolites: interactions of micorbes with sediments. Springer, Dordrecht
Tomlinson PB (2016) The botany of Mangroves. Cambridge University Press, Cambridge
Westerhold T, Marwan N, Drury AJ et al (2020) An astronomically dated record of earth's climate and its predictability over the last 66 million years. Science 369:1383–1387
Worthington TA, zu Ermgassen PSE, Fries DA et al (2020) A global biophysical typology of mangroves and its relevance for ecosystem structure and deforestation. Sci Rep 10:15652

The Caribbean Mangroves Today

2

Abstract

Extant mangroves cover a total of approximately 14,700 km^2 along Caribbean coasts, which represent ~10% of the total world's mangrove extent. The main components, in terms of fidelity, physiology, structure, and abundance (also known as major true mangrove elements of mangrove-forming trees), are *Rhizophora* (Rhizophoraceae), *Avicennia* (Acanthaceae), and *Laguncularia* (Combretaceae). Other minor true-mangrove elements are *Pelliciera* (Tetrameristaceae) and *Acrostichum* (Pteridaceae), which occupy marginal mangrove environments. A third category of ~20 mangrove associates are not restricted to mangrove communities and lack morphological and physiological adaptations to grow in saline waters. Relevant examples are the herb *Crenea* (Lythraceae) and the tree *Conocarpus* (Combretaceae). Biogeographically, the Caribbean mangroves are within the Atlantic-East Pacific (AEP) region, having a significantly poorer mangrove flora than the Indo-West Pacific (IWP) region, both separated by the continental African barrier. Caribbean mangroves typically develop sea–land zonation along salinity gradients governed by topography and tidal influence. The most common pattern, from sea to land, is *Rhizophora*-(*Pelliciera*)-*Avicennia*-*Laguncularia*-*Conocarpus*. Pollen and spores (sporomorphs) from true and associate mangrove elements are the most used evidence to reconstruct the ecological, biogeographical, and evolutionary trends of these communities. This book gathers the available sporomorph records (156 localities) in a dataset called CARMA (CARibbean MAngroves) as the basis for the following discussions.

To properly understand the origin and evolution of a given biome, vegetation type, or ecosystem, it is first necessary to be aware of the subject we are dealing with in its present-day manifestation, which is the one that we are able to perceive. This chapter briefly describes the main environmental traits of Caribbean mangroves and

V. Rull, *Origin and Evolution of Caribbean Mangroves*, Ecological Studies 252,
https://doi.org/10.1007/978-3-031-57612-6_2

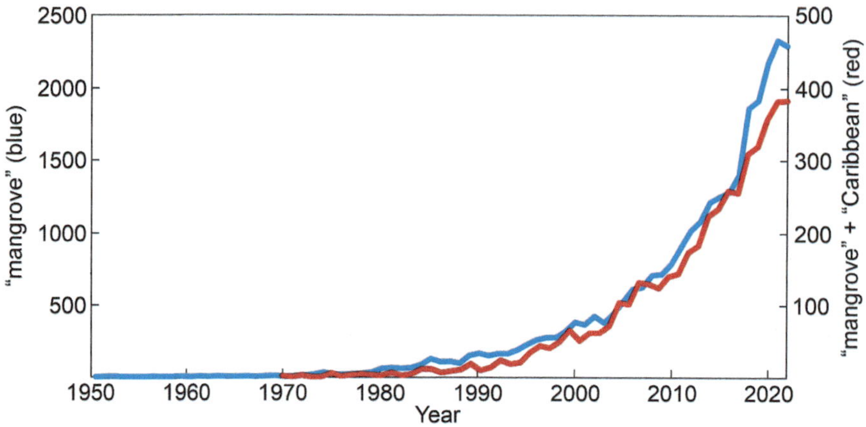

Fig. 2.1 Results of a Scopus search performed on January 6, 2023, using the words "mangrove" (blue line; 28,374 total records) and "mangrove" plus "Caribbean" (red line; 4638 total records)

the plant taxa that define them, along with their main biogeographical and ecological features. These aspects are treated in more depth in the specialized literature, which is mentioned in the text for those interested in more specific topics. Here, only a summary is provided to cover the aspects needed to understand the following chapters. Only plants are considered, as they are the fundamental mangrove builders and organizers and because of the personal author's bias, who uses plant fossils (mainly pollen) to reconstruct the ecological, biogeographical, and evolutionary history of mangroves. Other mangrove elements (animals, fungi, microorganisms, etc.) may be equally important for mangrove characterization and ecological dynamics, but they are left to be properly discussed by the corresponding experts. There is a large amount of literature that could be useful in this sense and would be impossible to summarize here. To provide an idea on the magnitude of the available literature, a recent Scopus search using the word "mangrove" produced almost 28,400 results, with only sporadic publications until the 1950s, a more consistent trend (up to 100 items per year) since the 1970s and an exponential increase (1000–2000 items per year) in the last decade (Fig. 2.1). Of these publications, >4600 (16%) also include the word "Caribbean" and show similar trends, beginning in the 1970s.

2.1 Main Environmental Traits

Before describing the main environmental features of the study area, it should be highlighted that this book considers the Caribbean region in a wide sense, as depicted in Fig. 1.1. A number of Caribbean studies encompass only the islands, whereas others also include the Caribbean coasts of the surrounding countries. Here, all islands and/or countries situated around the Caribbean Sea, from the Greater Antilles (north) to northern South America (south), and from Central

America (west) to the Lesser Antilles (east), are considered. This is mainly due to biogeographical and evolutionary reasons. On the one hand, this wide Caribbean region is a consistent biogeographical unit, which is especially true in the case of mangroves. On the other hand, Caribbean mangroves have evolved in a highly dynamic paleographical framework involving many areas that are no longer part of the present Caribbean coasts, but they were part of the proto-Caribbean coasts in the past and have been important for the shaping of extant mangrove communities.

2.1.1 Caribbean Climates

In tropical regions, such as the Caribbean, the yearly fluctuations in average temperatures are typically minor, ranging from 2 °C to 7 °C. On the other hand, rainfall exhibits significant variability in both space and time (Taylor and Alfaro 2005). In the Caribbean, the annual average air temperatures generally fall within the range of 22 °C to 28 °C, with the lowest temperatures occurring in December and January and the highest in the period from June to September. Sea surface temperatures (SSTs) typically range from 25 °C (December) to 30 °C in July (Mona 2020). The total annual precipitation varies considerably, ranging from less than 1000 mm/year to over 4000 mm/year. The lowest levels of precipitation are found in the Lesser Antilles and the northern coasts of South America, while the highest levels are observed in Central America and northwest South America, where the Chocó region—one of the most humid regions in the world, which receives as much as 13,000 mm of rainfall per year (Yepes et al. 2019) (Fig. 2.2)—lies. Over the course of the twentieth century, a consistent warming trend of approximately 0.1 °C per decade has been observed, coinciding with the global warming trend of the past century (Jones et al. 2015; IPCC 2019). In contrast, there have been no significant upward or downward trends in rainfall during the same period, with precipitation patterns largely influenced by interannual and decadal variations (Mona 2020; Jones et al. 2015).

The defining characteristic of Caribbean weather is the yearly pattern of rainfall, which displays a distinct bimodal structure featuring two peaks of approximately 150–250 mm per month in May–June and September–October, as well as a minimum of approximately 50 mm per month in December–April (Fig. 2.2). The wet season, spanning from May to October, experiences a relatively dry period in July and August known as the Mid-Summer Drought (MSD) (Magaña et al. 1999; Herrera et al. 2020). This marked seasonality results from the interplay of various large-scale climatic systems, notably the North Atlantic Subtropical High (NASH), also referred to as the Azores High (AH), the eastern Pacific Intertropical Convergence Zone (ITCZ), the Atlantic Warm Pool (AWP), and the Caribbean Low-Level Jet (CLLJ). The NASH and ITCZ play crucial roles in facilitating moisture, with their seasonal movements, expansions, and contractions serving as the primary drivers of the bimodal precipitation pattern. NASH primarily affects the central and eastern Caribbean islands (zone c in Fig. 2.2), while the ITCZ exerts more influence on the western sector (zone b in Fig. 2.2). The AWP and CLLJ modulate the amount of moisture provided by the NASH and ITCZ as they undergo seasonal changes (Martínez et al. 2019).

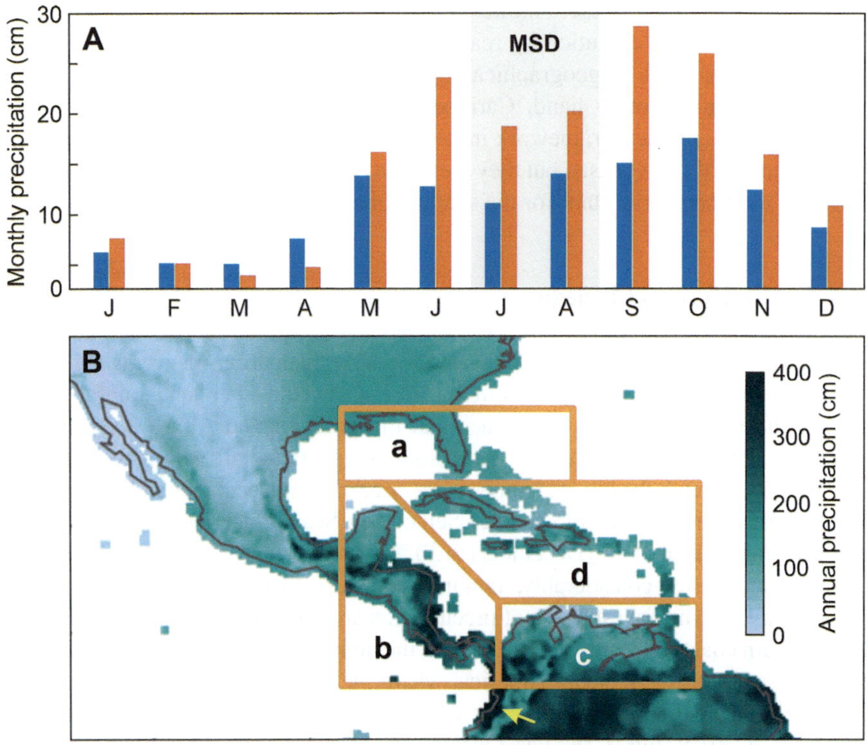

Fig. 2.2 Precipitation patterns across the Caribbean region (1950–2018 averages). (**a**) Annual precipitation cycle in Central America (orange) and the Caribbean islands (blue). The Mid-Summer Drought (MSD) is highlighted in gray. (**b**) Annual mean precipitation and subregional subdivision (a, Florida; b, Central America; c, northern South America; d, Caribbean islands). The Chocó region is indicated by a yellow arrow. Modified from Herrera et al. (2020)

2.1.2 Meteorological Hazards

The Caribbean region is susceptible to natural disasters driven by climate, resulting in substantial economic losses. Major threats include tropical cyclones (TCs), also known as hurricanes, and the associated heavy rainfall, floods, and landslides (Palmieri et al. 2006; Burgess et al. 2018). The hurricane season lasts from June to November, coinciding with weaker North Atlantic easterly trade winds and SSTs above 26 °C, creating optimal conditions for cyclonic activity (Mona 2020). Since 1995, there has been a significant increase in both hurricane activity and the prevalence of intense storms (Webster et al. 2005). The El Niño-Southern Oscillation (ENSO) plays a crucial role in modifying hurricane activity, either increasing it during La Niña conditions or reducing it during El Niño conditions, affecting the number and intensity of these events.

The Caribbean region, particularly its islands, experiences brief yet severe droughts that result in significant economic losses. These droughts impact key sectors such as agriculture (which is the cornerstone of the regional economy), water supply, hydroelectric power production, and tourism (Pielke et al. in 2003; Durán-Quesada et al. 2020). The occurrence and intensity of droughts exhibit notable variations across different locations and time periods, making it challenging to identify consistent patterns for prediction. For instance, various parts of the Caribbean have distinct drought cycles, ranging from 2–3 years to 9–13 years. Despite this variability, there appears to be a potential association with El Niño events, as indicated by correlations with SSTs in the tropical and equatorial Pacific. However, it remains difficult to establish clear long-term trends, as some Caribbean regions are becoming drier while others are showing a tendency toward more humid conditions (Mona 2020).

Generally, from 1960 to 2013, across 22 Caribbean nations, 370 meteorological disasters were reported. Among these, TCs were the most common, making up approximately 72% of the total, with riverine flooding being the next most prevalent at approximately 16%. Moreover, during this same time frame, TCs were responsible for a staggering 95% of the economic losses incurred. It is worth noting that the occurrence of meteorologically linked natural disasters has seen a substantial increase, going from less than 2 events per year prior to 1980 to more than 10 events per year thereafter (Burgess et al. 2018).

2.1.3 Sea Level

According to Church and White (2006), global sea-level rise rates have increased from 1.7 mm/year in the twentieth century to 3.2 mm/year since the beginning of the current century. In the Caribbean Sea, measurements indicate rising rates of 1.8 mm/year from 1950 to 2009 and 2.5 mm/year from 1993 to 2010 (Palanisamy et al. 2012; Torres and Tsimplis 2013). However, it is important to note that there are important deviations from these values, revealing significant spatial variation. The extremes range from 0.3 mm/year on the northern coasts of Venezuela in South America to 10.8 mm/year in Haiti (Hispaniola). These regional disparities are attributed to factors such as postglacial rebound and local tectonics.

2.1.4 Future Projections

The current projections for climate change in the Caribbean region indicate that temperatures are expected to rise between 1 and 3.5 °C by the end of the century, affecting both the land and sea. The most significant temperature increases are anticipated on land, and there will be a higher occurrence of extreme temperature events. These models also suggest a moderate decrease in rainfall, with a reduction of 20–30%, as well as an increase in heavy rainfall events and more consecutive hot and dry days (Campbell et al. 2011; Stennett-Brown et al. 2017; Taylor et al. 2018). The frequency and intensity of storms and hurricanes are also expected to increase in line with global projections. In terms of sea levels, the Caribbean forecasts coincide with global estimates, indicating a rise of 0.4–0.6 m relative to the period between 1986 and 2005, although some studies caution that increases beyond 1 m should not be ruled out (Rahmstorf 2007; Perrette et al. 2013).

In addition to a wide range of human activities, these occurrences can have an impact on both terrestrial and marine ecosystems, as well as biodiversity. For instance, rising temperatures can lead to coral bleaching, encourage the proliferation of invasive species, and support the northward migration of Caribbean fish and coral reefs. Reduced rainfall can influence the patterns of seed production, germination, survival, and growth, ultimately increasing seedling mortality. The increased frequency of floods, storms, and hurricanes can exacerbate the degradation of coastal wetlands and forests, particularly mangroves, diminishing their natural filtration and buffering capabilities. This can exacerbate habitat loss and contribute to the deterioration of other ecosystems, such as coral reefs, ultimately leading to a decline in biodiversity. Sea-level rise can also impact mangroves by prompting them to migrate inland and intensifying coastal erosion (Mona 2020).

2.2 Mangrove Plant Communities

2.2.1 Extent

According to the most recent estimates, corresponding to the year 2020 (Bunting et al. 2022), Caribbean mangroves have a total extent of approximately 14,700 km^2, which represents ~10% of the total world's mangrove extent (Fig. 2.3). The countries with more extensive mangrove cover are Cuba, Venezuela, Colombia, and Panama (1500–3600 km^2); all other countries are at or below 600 km^2, and 15 of them have less than 100 km^2 of mangroves (Table 2.1). In the recent decades, Caribbean mangroves have been reduced by almost ~500 km^2 (3.3%, very close to the global value) due to natural and anthropogenic deforestation. This has been mainly due to significant reductions (2–8%) in countries with >100 km^2 of mangroves, with the exceptions of Guatemala and Venezuela, where mangroves decreased less than 1%. The overall reduction rate is ~21 km^2 per year. At these rates, the disappearance of all Caribbean mangroves may be a matter of centuries (Rull 2023b). We return to this topic in Chap. 8, when discussing mangrove conservation.

2.2.2 Floristics

In terms of plant composition within mangrove ecosystems, there are two primary categories: true (strict) mangrove elements and mangrove associates (Table 2.2). To be classified as a true mangrove element, the following criteria must be met (Tomlinson 2016):

(i) Exclusive to mangroves, without presence in terrestrial habitats
(ii) Plays a significant role in shaping the community and can form pure stands
(iii) Displays specific morphological adaptations to thrive in intertidal conditions, typically pneumatophores and viviparous embryos

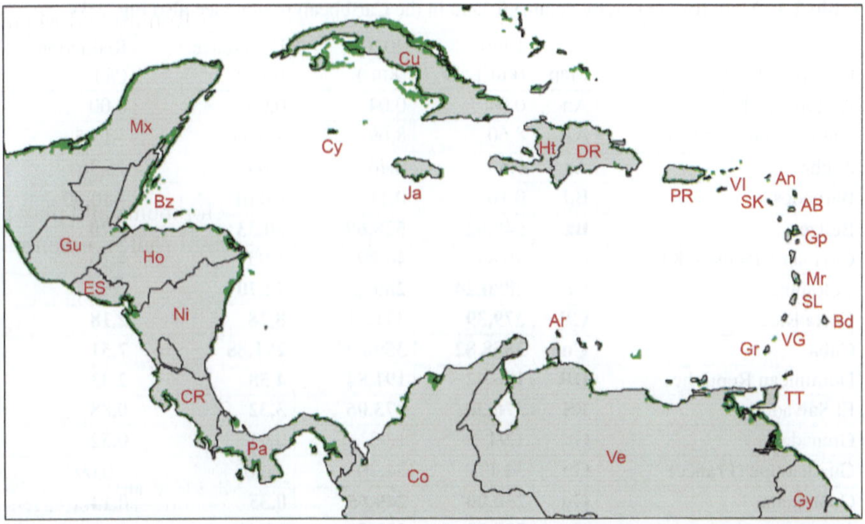

Fig. 2.3 Extent of the Caribbean mangroves (green patches), according to Bunting et al. (2022). Base image: NASA Landsat 5-TM (https://earthobservatory.nasa.gov/images/47427/mapping-mangroves-by-satellite). Country/island abbreviations as in Table 2.1

(iv) Possesses physiological mechanisms for excluding salt, as an adaptation for growth in saline waters

(v) Demonstrates systematic isolation from its terrestrial relatives, typically at the generic level and sometimes at the family or subfamily level

True mangrove elements can be further categorized as major or minor. Major true mangrove elements, often referred to as mangrove-forming trees, are primarily trees and belong to the genera *Rhizophora* (Rhizophoraceae), *Avicennia* (Acanthaceae), and *Laguncularia* (Combretaceae), as mentioned in the introduction. On the other hand, minor true mangrove elements share similar characteristics but do not play a significant structural role in the community (condition ii) and cannot form pure stands (condition iii). They are typically found in peripheral intertidal areas. Examples of minor elements include species from *Pelliciera* (Tetrameristaceae) and *Acrostichum* (Pteridaceae), with *Pelliciera* sometimes forming small pure stands under extremely humid and shaded conditions (Dangremond et al. 2015). Mangrove associates are plants that are commonly found in mangrove environments but are not limited to them (condition i). They do not have a substantial structural impact on the ecosystem (condition ii) and lack specific morphological and physiological adaptations for intertidal life (conditions iii and iv). These associates can also be found in various other habitats such as coastal swamps, back-mangrove wetlands, salt marshes, riverbanks, beach communities, and inland rainforests (Tomlinson 2016). For instance, the herb *Crenea maritima* (Lythraceae) is unique to mangrove environments (condition i) but does not meet conditions ii and iii; and therefore, it is categorized as a mangrove associate. *Conocarpus erectus*, although capable of forming self-sustaining stands (condition ii), lacks morphological adaptations (condition iii) and cannot endure flooding and high salinity (condition iv), and it is typically found in marginal mangrove environments (Lonard et al. 2020b).

Table 2.1 Mangrove extent by country/island in the Caribbean region (see also Fig. 2.3)

Country/island	Map	1996 (km²)	2020 (km²)	Difference (km²)	Reduction (%)
Anguilla (UK)	An	0,04	0,04	0,00	0,00
Antigua and Barbuda	AB	8,60	8,69	−0,09	−1,05
Aruba	Ar	0,55	0,46	0,09	16,36
Barbados	Bd	0,10	0,11	−0,01	−10,00
Belize	**Bz**	**549,02**	**528,69**	**20,33**	**3,70**
Cayman Islands (UK)	Cy	46,84	44,89	1,95	4,16
Colombia	**Co**	**2880,24**	**2807,54**	**72,70**	**2,52**
Costa Rica	**CR**	**379,39**	**371,11**	**8,28**	**2,18**
Cuba	**Cu**	**3888,82**	**3596,94**	**291,88**	**7,51**
Dominican Republic	**DR**	**196,42**	**191,84**	**4,58**	**2,33**
El Salvador	**ES**	**376,38**	**373,06**	**3,32**	**0,88**
Grenada	Gr	1,94	1,93	0,01	0,52
Guadeloupe (France)	Gp	34,17	34,20	−0,03	−0,09
Guatemala	**Gu**	**250,00**	**249,65**	**0,35**	**0,14**
Guyana	**Gy**	**308,08**	**288,59**	**19,49**	**6,33**
Haiti	**Ht**	**166,81**	**154,01**	**12,80**	**7,67**
Honduras	**Ho**	**624,17**	**605,64**	**18,53**	**2,97**
Jamaica	Ja	105,51	99,45	6,06	5,74
Martinique (France)	Mr	19,21	19,41	−0,20	−1,04
Nicaragüa	**Ni**	**763,04**	**747,31**	**15,73**	**2,06**
Panama	**Pa**	**1558,22**	**1535,69**	**22,53**	**1,45**
Puerto Rico	PR	85,50	82,84	2,66	3,11
Saint Kitts and Nevis	SK	0,34	0,35	−0,01	−2,94
Saint Lucia	SL	1,63	1,62	0,01	0,61
Saint Vincent and The Grenadines	VG	0,32	0,33	−0,01	−3,13
Trinidad and Tobago	TT	83,07	82,23	0,84	1,01
Venezuela	**Ve**	**2848,56**	**2846,75**	**1,81**	**0,06**
Virgin Islands (UK/USA)	VI	3,57	3,59	−0,02	−0,56
Total		**15,180,54**	**14,676,96**	**503,58**	**3,32**

Countries/islands with >100 km² of total mangrove extent are in bold. Raw data from Bunting et al. (2022)

Apart from the aforementioned true and associated mangrove species, approximately 120 other accompanying species have been documented in Neotropical mangroves (Appendix 1). These species define 30 distinct phytosociological associations, all of which are found in the Caribbean region (Table 2.3). These associations are primarily characterized by the dominance or subdominance of species belonging to the genera *Rhizophora*, *Avicennia*, *Laguncularia*, *Conocarpus*, and *Acrostichum*. Additionally, there are 85 species unique to one group or another, serving as diagnostic species in this context. Group A includes associations where the primary mangrove-forming trees are *A. germinans* and *L. racemosa*, with limited representation of *R. mangle*. In group B, associations are dominated by either *C. erectus* (B_1) or *A. germinans* (B_2). The most significant tree in group C is *R. mangle*, either on its own (C_1) or in conjunction with *L. racemosa* and, in one instance, with *P. rhizophorae* (tea mangrove) (C_2).

Table 2.2 True (major and minor) and associate mangrove plant elements of the Caribbean region

Type		Species	Family	Plant type
True	Major	*Avicennia bicolor* Standl.[a]	Acanthaceae	Tree
		Avicennia germinans (L.) Stearn[a]	Acanthaceae	Tree
		Avicennia schaueriana Stapf & Leechm. ex Moldenke[a]	Acanthaceae	Tree
		Laguncularia racemosa C.F.Gaertn.[a]	Combretaceae	Tree
		Rhizophora mangle L.[a]	Rhizophoraceae	Tree
		Rhizophora racemosa (G.Mey.) Engl.[a]	Rhizophoraceae	Tree
	Minor	*Acrostichum aureum* L.	Pteridaceae	Fern
		Acrostichum daneaeifolium Langsd. & Fisch.[a]	Pteridaceae	Fern
		Pelliciera benthamii (Planch. & Triana) N.C.Duke	Tetrameristaceae	Tree
		Pelliciera rhizophorae Planch. & Triana[a]	Tetrameristaceae	Tree
Associate		*Amphitecna latifolia* (Mill.) A.H.Gentry	Bignoniaceae	Tree
		Anemopaegma chrysoleucum (Kunth) Sandwith	Bignoniaceae	Vine
		Batis maritima L.	Batidaceae	Shrub
		Caesalpinia bonduc (L.) Roxb.	Fabaceae	Tree
		Conocarpus erectus L.[a]	Combretaceae	Tree
		Crenea patentinervis (Koehne) Standl.[a]	Lythraceae	Herb
		Dalbergia ecastaphyllum Taub.	Fabaceae	Tree/Shrub
		Dalbergia amerimnum Benth.	Fabaceae	Tree/Shrub
		Hibiscus tiliaceus L.	Malvaceae	Tree
		Hippomane mancinella L.	Euphorbiaceae	Tree
		Mora oleifera Duke[a]	Fabaceae	Tree
		Muellera moniliformis L.f.[a]	Fabaceae	Tree
		Pachira aquatica Aubl.	Bombacaceae	Tree
		Pavonia rhizophorae Killip[a]	Malvaceae	Shrub
		Pavonia spicata Cav.	Malvaceae	Shrub
		Phryganocydia phellosperma (Hemsl.) Sandwith	Bignoniaceae	Vine
		Pluchea odorata (L.) Cass.	Asteraceae	Herb
		Rhabdadenia biflora Müll.Arg.	Apocynaceae	Vine
		Rustia occidentalis (Benth.) Hemsl.	Rubiaceae	Tree/Shrub
		Scaevola plumieri (L.) Vahl	Goodeniaceae	Shrub
		Tabebuia palustris Hemsl.[a]	Bignoniaceae	Tree
		Thespesia populnea (L.) Sol. ex Corrêa	Malvaceae	Tree
		Thespesia populneoides (Roxb.) Kostel.	Malvaceae	Tree
		Tuberostylis axilaris S.F.Blake	Asteraceae	Shrub
		Tuberostylis rhizophorae Steetz	Asteraceae	Epiphyte

Based on Gentry (1982), Tomlinson (2016), and Duke (2017). Nomenclature according to the International Plant Names Index (IPNI) (https://www.ipni.org/)

[a]Species used by Duke (2017) to characterize the Atlantic-East Pacific (AEP) mangroves (Fig. 2.4)

Table 2.3 Phytosociological groups, subgroups, and associations defined for Neotropical mangrove formations (García-Fuentes et al. 2020)

Group	Association	Dominant	Subdominant	Exclusive species (Appendix 1)
A	Lag-AviP	*A. germinans*	*L. racemosa*	*S. pacifica*
	Lag-AviR	*A. germinans* *L. racemosa*	*C. erectus*	–
	Lon-LacC	*L. racemosa*	*A. germinans* *C. erectus*	*L. sericeus, M. pigra, P. elsiae*
	Sta-LagC	*L. racemosa*	*A. germinans*	–
	LagracR	*L. racemosa*	–	–
B_1	Acr-ConB	*C. erectus* *A. aureum*	*L. racemosa*	*C. mariscus, A. daneaeifolium, E. cellulosa, F. spadicea, I. triloba, S. americanus, H. plaeana, Q. odoratissima, A. macracantha*
	Lon-ConC	*C. erectus*	*R. mangle* *L. racemosa* *A. germinans*	*L. pycnophyllus*
	Spo-ConM	*C. erectus*	–	*J. keyensis, P. royenii, P. mucronatum, T. recurvata*
	Dis-ConR	*C. erectus*	–	*P. distachyon, C. elara*
	Bat-ConM	*C. erectus*	–	*E. convolvuloides, F. cimosa, S. grandiflorus, B. lindenii, C. yucatanensis, S. linearis, A. hispida*
	Rac-ConM	*C. erectus*	–	–
	ConereR	*C. erectus*	–	*L. involucrata, C. clusiifolia, E. littoralis*
	Con-CocR	*C. erectus*	–	*T. populnea*
	Gra-ConC	*C. erectus*	*L. racemosa* *A. germinans*	*P. acapulcensis, P. quadrangulare*
	Lag-ConP	*C. erectus*	*L. racemosa*	–
B_2	Lyc-ConR	*A. germinans* *C. erectus*	–	*C. barbata, C. verticillata, H. eriophora, D. virgatus, L. carolinianum, O. dillenii*
	Sar-AviM	*A. germinans*	*L. racemosa*	–
	Bat-AviB	*A. germinans*	*C. erectus*	*S. pyramidatus, H. curassavicum, I. cheiranthifolia*
	AvigerR	*A. germinans*	–	*B. viscosa*
	Bat-AviP	*A. germinans*	–	*M. littoralis*
	Avi-RhiR	*A. germinans* *R. mangle*	*C. erectus*	–
C_1	Pri-RhiC	*R. mangle*	*A. aureum*	*P. copaifera, B. guineensis, C. melo, I. alba, B. hyacinthina, C. cujete, D. beownei, M. pendula*
	Avi-RhiB	*R. mangle*	–	*A. schaueriana*
	Ann-FicC	*R. mangle*	–	*C. aculeata, E. fusca, K. anomala*
	RhimanC	*R. mangle*	–	–

(continued)

Table 2.3 (continued)

Group	Association	Dominant	Subdominant	Exclusive species (Appendix 1)
C_2	Pel-RhiC	*R. mangle* *L. racemosa* *P. rhhizophorae*	*A. germinans*	–
	Rha-LagC	*L. racemosa*	*R. mangle* *A. germinans*	*C. alternifolius, E. polystachia, E. interstincta, H. ritteriana, E. mutata, L. octovalvis, P. acuminata*
	Mac-RhiC	*R. mangle*	*L. racemosa* *A. germinans*	*N. multiflora, D. berteroi, S. causiarum*
	Dai-RhiB	*R. mangle*	*L. racemosa*	*B. palustris, H. penduliflora, T. usneoides, S. palmetto, R. regia, C. trifoliata, T. fasciculata, T. angustata*
	Lag-RhiP	*R. mangle* *L. racemosa*	–	*P. fuscescens, B. mucronatum, A. saman, E. prostrata, F. bullenei, G. sepium, M. tinctoria, M. trifidus, S. jamaicense, T. catappa, S. glaucum, W. fruticosa*

2.2.3 Biogeography

The distribution of mangrove species around the world is notably uneven, a phenomenon sometimes referred to as the mangrove "anomaly." This phenomenon gives rise to two main biogeographical regions: the Atlantic-East Pacific (AEP) region and the Indo-West Pacific (IWP) region (Fig. 2.4). Of the 69 mangrove species that can be found globally (belonging to 32 genera and 18 families), only 17 species (from 11 genera and 9 families) are found in the AEP region, while the remaining 54 species (from 24 genera and 17 families), which is over three times the number, are characteristic of the IWP region (Duke 2017). The specific AEP species present in the Caribbean are detailed in Table 2.2.

Only one species, *Acrostichum aureum*, is found worldwide, while the other species within this genus are limited to either the AEP (*A. danaeifolium*) or the IWP (*A. speciosum*) regions (Fig. 2.5). Two genera, *Rhizophora* and *Avicennia*, are shared between both the AEP and IWP regions, but they are represented by different species. The remaining relevant Caribbean elements (*Pelliciera*, *Laguncularia*, and *Conocarpus*) are exclusive to the AEP region, with *Pelliciera* being confined to the Neotropics, specifically within a relatively small equatorial area around the Panama Isthmus (Duke 2020). The genera exclusive to IWP mangroves include *Nypa* (Arecaceae); *Camptostemon* (Fabaceae); *Lumnitzera* (Combretaceae); *Sonneratia* (Lythraceae); *Aegiceras* (Myrsinaceae); *Osbornia* (Myrtaceae); *Aegialitis* (Plumbaginaceae); *Bruguiera*, *Ceriops*, and *Kandelia* (Rhizophoraceae); and *Scyphiphora* (Rubiaceae) (Duke in 2017).

Primary Caribbean mangrove species (*R. mangle*, *A. germinans*, *L. racemosa*, *C. erectus*, and *A. aureum/A. danaeifolium*) are prevalent throughout the entire region examined in this study. On the other hand, certain *Rhizophora* and *Avicennia*

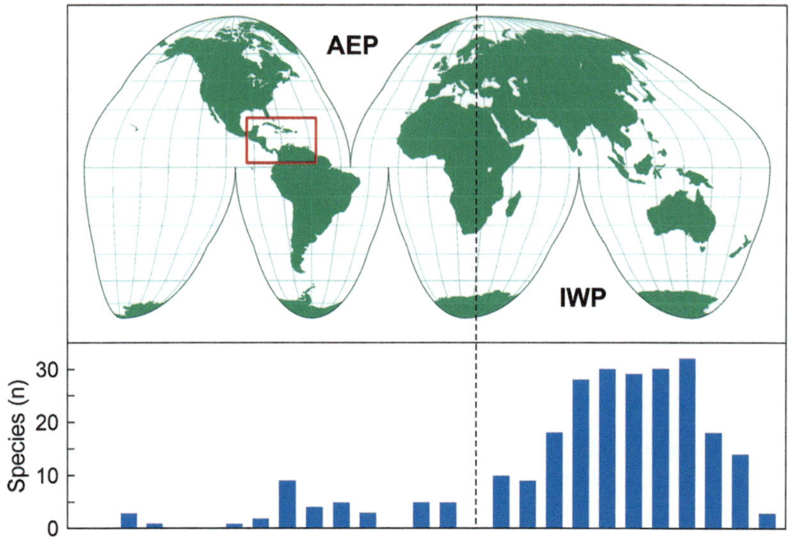

Fig. 2.4 Worldwide mangrove "anomaly" in species richness. The Caribbean region is marked by a red box. AEP, Atlantic-East Pacific; IWP, Indo-West Pacific. Based on Ellison et al. (1999)

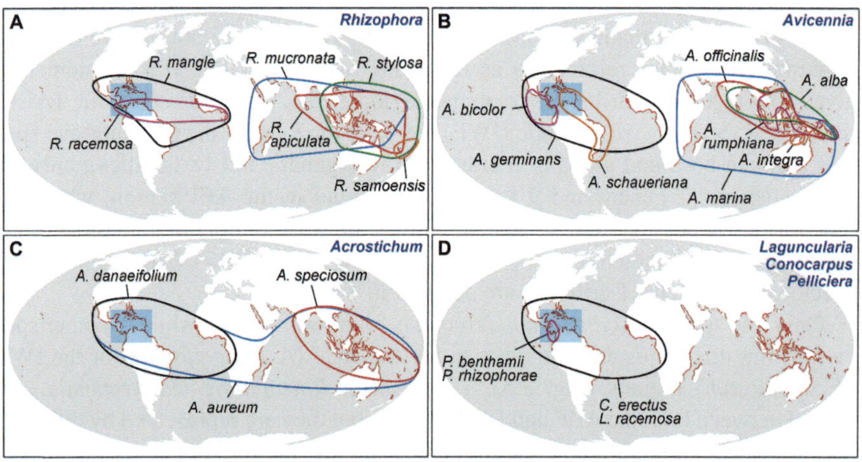

Fig. 2.5 Distribution of the Caribbean true-mangrove genera and their species (Table 2.2). (**a**) *Rhizophora* species. (**b**) *Avicennia* species. (**c**) *Acrostichum* species. (**d**) *Laguncularia*, *Conocarpus*, and *Pelliciera* species. The Caribbean region is highlighted by a transparent blue box. Based on Lo et al. (2014), Li et al. (2016), and Duke (2017)

species, such as *R. racemosa*, *A. bicolor*, and *A. schaueriana*, have more limited distributions (Fig. 2.6). Notably, *P. rhizophorae* stands out because its current distribution area is considered a remnant of the former range of *Pelliciera*, which was once widespread across the entire Neotropics during the Miocene. Various factors

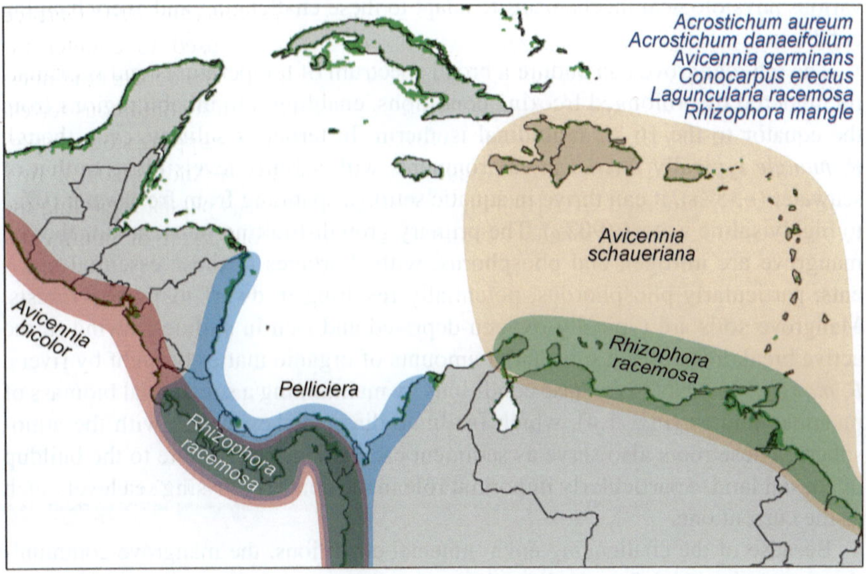

Fig. 2.6 Intra-Caribbean geographical ranges of true mangrove species (including *Conocarpus*) (Table 2.2). The species indicated in blue in the upper right corner are distributed across the whole Caribbean region. Based on Lo et al. (2014), Li et al. (2016), and Duke (2017, 2020)

contributing to this significant reduction have been debated, including low tolerance to environmental stressors (e.g., light, temperature, salinity, sea-level fluctuations) and competition with *Rhizophora* (Graham 1977, 1995). It has also been suggested that this relictualization might represent the final phase of an extended taxon cycle, potentially signaling the species' impending extinction (Rull 2023a). A more in-depth discussion of this topic can be found in Chap. 6.

2.2.4 Ecological Hints

In the Caribbean, *R. mangle*, also known as the red mangrove, stands as the dominant and most widely distributed tree responsible for forming mangrove habitats. A recent comprehensive examination of this species was conducted by DeYoe et al. (2020), and the following summary is based on their findings. *R. mangle* typically thrives in the intertidal area along the seaward edge of mangrove ecosystems. However, it can also be spotted at estuary entrances, along the banks of tidal creeks, and in the fringes of upper regions. This tree prefers soft mud or fine silt clay as its substrate, particularly in calm coastal areas where strong waves and significant tidal fluctuations are absent and erosion is not a significant issue. Despite the stability of the substrate, mangrove environments are often characterized by physiological stress factors such as varying salinity and nutrient availability, elevated temperatures, intense light exposure, and oxygen-depleted sediments. *R. mangle* has evolved various physiological mechanisms to adapt to these challenging and ever-changing conditions.

The red mangrove can endure a broad spectrum of temperatures and is primarily hindered by prolonged freezing conditions, enabling it to inhabit regions from the equator to the 16 °C latitudinal isotherm. In terms of salinity, even though *R. mangle* typically thrives in environments with salinity levels close to that of seawater (~35‰), it can thrive in aquatic settings spanning from freshwater (0‰) to highly saline waters (90‰). The primary growth-limiting nutrients for the red mangrove are nitrogen and phosphorus, with shortages in these essential nutrients, particularly phosphorous, potentially resulting in dwarf mangrove forests. Mangrove soils are typically oxygen-deprived and rich in sulfides, owing to the active breakdown of the substantial amounts of organic matter brought by rivers. *R. mangle* has adapted to these conditions by maintaining a substantial biomass of pneumatophores (Fig. 1.4), which facilitate direct gas exchange with the atmosphere. These roots also serve as sediment catchers and contribute to the buildup of coastal land, a particularly important role in the context of rising sea levels such as the current one.

Because of the challenging environmental conditions, the mangrove communities centered around *R. mangle*, and associated tree species have limited biodiversity, and the understory beneath the forest canopy is poorly developed, consisting of a small number of herbs, shrubs, and vines. Specific land-sea community zonation arises due to the interaction of various biotic and abiotic factors. Notably, factors such as topography, tidal influence, and the resulting salinity gradient play a significant role among the abiotic factors. The competition between different mangrove species and the dispersal of seeds are the primary biotic factors used to explain the distribution of mangroves (Lugo and Snedaker 1974; Rabinowitz 1978; Woodroffe 1982; Smith 1992; Sousa et al. 2007). The most common pattern of mangrove zonation, starting from the outermost landward edge, includes *R. mangle*, *A. germinans*, *L. racemosa*, and *C. erectus*, overlapping with brackish-water swamp communities further inland, followed by freshwater marsh areas and inland rainforests and savannas (Fig. 2.7). When *P. rhizophorae* is present, it thrives near the outer fringe within the *Rhizophora* canopy, which provides protection against intense light and salinity conditions (Dangremond et al. 2015). The zonal pattern may exhibit minor variations due to local differences in the influencing factors. In some cases, slight shifts in the positions of *A. germinans* and *L. racemosa* along the gradient occur (Lonard et al. 2017, 2020a). In other situations, limited environmental variability and the absence of permanent freshwater sources can lead to a lack of any discernible zonal pattern, resulting in a nearly uniform composition of mangroves (Urrego et al. 2009).

The black mangrove, *A. germinans*, is capable of thriving in both fresh and highly saline waters and possesses distinct pneumatophores (Fig. 1.4). This species serves as a pioneering mangrove in the upper intertidal zone and can establish itself on unstable mud banks, forming temporary monospecific communities without any understory vegetation. It is important to note that the dispersal of black mangrove propagules is unsuccessful in consistently flooded or completely dry soils. Instead, their seedlings flourish in locations where the water level is close to the surface and

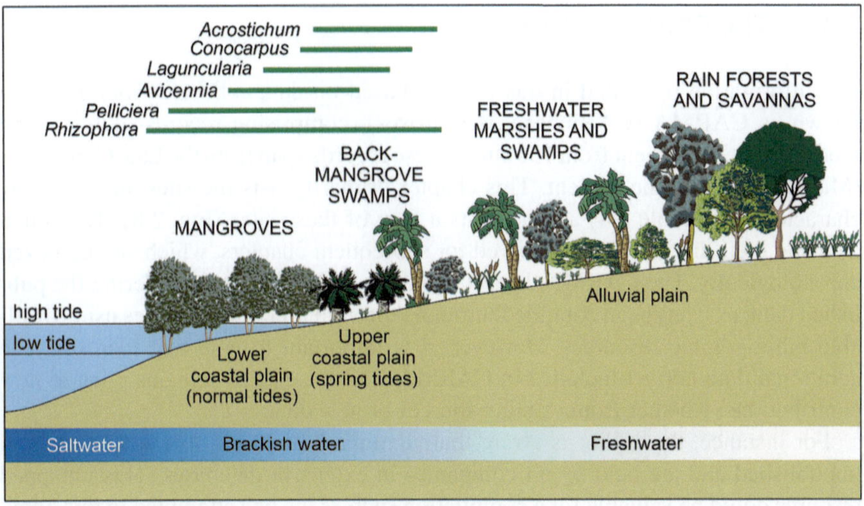

Fig. 2.7 Schematic transect illustrating the characteristic coastal zonation of Caribbean mangroves. Green lines represent the approximate ranges of the most important elements. After Rull (2022)

the salinity is relatively low (Lonard et al. 2017). In contrast, the white mangrove, *L. racemosa*, seldom dominates mangrove stands. It occupies a specific ecological niche on nutrient-rich elevated soils, which experience moderate to infrequent tidal flooding and frequent freshwater inflows, typically under intermediate salinity conditions ranging from 15 to 20‰ (Lonard et al. 2020a, b). Outside of these conditions, the species struggles to compete with other mangrove trees, especially *Rhizophora*. In some instances, the dominance of *L. racemosa* in certain mangrove communities is linked to frequent disturbances and elevated nutrient concentrations, particularly phosphorus and nitrogen (Medina et al. 2015). It is important to note that *Laguncularia* has lower cold tolerance than *Rhizophora* and *Avicennia*, which restricts its geographical distribution.

In landward settings, mangrove species such as *Conocarpus* (buttonwood) and *Acrostichum* (mangrove fern) are commonly found. *C. erectus* is typically found in elevated areas in the transition zone between mangroves and upland tropical forests. It can tolerate salinity levels of up to 10‰ and does not develop pneumatophores (Fig. 1.4). This species cannot survive in flooded soils but is known for its ability to withstand high temperatures (Lonard et al. 2020a, b). On the other hand, *Acrostichum* can thrive in a broad range of salinity levels but typically prefers brackish waters. Although it can tolerate shade, its optimal growth conditions require full sun exposure (Medina et al. 1990). These specific environmental conditions are often found in back-mangrove areas (Fig. 2.7) and in disturbed areas where the fern takes advantage of canopy gaps. In such situations, it can impede forest regeneration and become one of the most aggressive invaders of mangrove ecosystems following both natural and human-induced disturbances (Blanco et al. 2012; Biswas et al. 2018).

2.3 The CARMA Dataset

The information presented in this book is based on data collected from a dataset known as CARMA (CARibbean MAngroves), comprising records of fossil and modern pollen sediment from 156 sites. These records span from the Late Cretaceous (Maastrichtian) to the present. This chapter primarily lists the sites and their key characteristics (Table 2.4) and presents a map of these sites (Fig. 2.8). The actual data and their discussion are reserved for subsequent chapters, which are organized chronologically. The CARMA dataset is quite comprehensive considering the published data, as it has been compiled through thorough literature searches using available bibliographic resources. However, it is important to note that there may be additional data not yet included in CARMA, and the dataset remains open to new contributions, whether from existing sources or new data.

For instance, the author is aware that significant amounts of raw data are still unpublished and are held by oil companies in extensive databases. This untapped resource could be valuable for a compilation such as the one attempted in this book. Thankfully, some industrial biostratigraphers have been able to publish pertinent information in this context. Notable syntheses by authors such as Kuyl et al. (1955), Van der Hammen and Wijmstra (1964), Wijmstra (1968, 1969), Germeraad et al. (1968), Muller (1959, 1981), Lorente (1986), and Muller et al. (1987) have significantly contributed to our understanding of Neotropical plant evolution. However, it is evident that renewed efforts are necessary to access classified data for the advancement of evolutionary knowledge. This endeavor can be pursued without violating the confidentiality rules of the involved companies, as exemplified by the authors mentioned earlier and others. The CARMA dataset, encompassing both raw data and graphical representations, has been reviewed and updated for this book, with some minor errors corrected. The version utilized in this book is considered the most comprehensive and reliable based on the available information. However, periodical updates will be conducted and uploaded to public repositories (Rull 2023c).

In its current form, the CARMA dataset consists of 5 locations (3%) representing the Late Cretaceous, 35 (23%) from the Paleogene, 43 (27%) from the Neogene, 60 (39%) from the Quaternary, and 12 (8%) from modern sediments. The primary methods used for dating are stratigraphic/biostratigraphic correlation for rocks predating the Quaternary period and predominantly radiometric techniques, with radiocarbon dating being the most commonly employed for Quaternary sediments. The main sources for Quaternary data are sediment cores, primarily from lakes and coastal peat bogs, while Late Cretaceous to Neogene records are primarily derived from well samples, with a smaller portion coming from outcrops. Notably, there is a lack of deep-sea core data for mangrove studies. All Quaternary records are quantitative, and nearly one-fifth of pre-Quaternary records are either qualitative or semiquantitative, resulting in nearly 90% of the studies being quantitative. The most frequently used proxy is pollen, employed in 130 (92%) of the cases.

In terms of location, the dataset displays a notably higher concentration of sites in the southern Caribbean, particularly in northern South America (Fig. 2.8). This is primarily because of the prevalence of research conducted in Colombia and

Table 2.4 Caribbean localities with Late Cretaceous to present mangrove records

N	Site/area	Map	Country	ArXiv	Age	Dating method	Proxies	Data	References
1	Cayo Coco	CC	Cuba	Surface sed	Modern	–	P	Qt	Davidson (2007)
2	Changuinola	Cg	Panama	Surface sed	Modern	–	P	Qt	Phillips et al. (1997)
3	Coastal Plain	CP	Guatemala	Surface sed	Modern	–	P	Qt	Van der Hammen (1963)
4	Cispatá	Cs	Colombia	Surface sed	Modern	–	P	Qt	Urrego et al. (2009)
5	Lake Maracaibo	LM	Venezuela	Surface sed	Modern	–	P	Qt	Tschudy (1969)
6	Caño Macareo	CM	Venezuela	Surface sed	Modern	–	P	Qt	Hofman (2002)
7	Playa Medina	Me	Venezuela	Surface sed	Modern	–	P	Qt	Rull and Vegas-Vilarrúbia (1999)
8	NW CR	NC	Costa Rica	Surface sed	Modern	–	P	Qt	Rodgers and Horn (1996)
9	Orinoco delta	OD	Venezuela	Surface sed	Modern	–	P	Qt	Muller (1959)
10	Los Petenes	Pe	Mexico	Surface sed	Modern	–	P	Qt	Escarraga-Paredes et al. (2014)
11	San Andrés	SA	Colombia	Surface sed	Modern	–	P	Qt	Urrego et al. (2010)
12	Caño Tigre	Ti	Venezuela	Surface sed	Modern	–	P	Qt	Montoya et al. (2019)
13	Candelaria	Cd	Colombia	Sed core	L Holocene	ND	P	Qt	Urrego et al. (2018)
14	Uno	Un	Colombia	Sed core	L Holocene	ND	P	Qt	Urrego et al. (2018)
15	Navío	Nv	Colombia	Sed core	L Holocene	^{210}Pb	P/G	Qt	Castaño et al. (2010)
16	Espíritu Santo	ES	Puerto Rico	Sed core	L Holocene	$^{14}C/^{137}Cs$	G/F/T	Qt	Khan et al. (2019)
17	Rosario	Ro	Colombia	Sed core	L Holocene	$^{137}C/^{210}Pb$	P	Qt	Urrego et al. (2019)
18	Flotante	LF	Colombia	Sed core	L Holocene	^{210}Pb	P/G	Qt	Castaño et al. (2010)
19	Alejandro	Aj	Dominican Rep	Sed core	L Holocene	^{14}C	P/Ch/G/Py	Qt	LeBlanc et al. (2017)
20	Honda	Hd	Colombia	Sed core	L Holocene	^{14}C	P	Qt	González et al. (2010)
21	Manatee	Ma	Jamaica	Sed core	L Holocene	^{14}C	O/G	Qt	Palmer et al. (2020)
22	Calancala	Cl	Colombia	Sed core	L Holocene	$^{14}C/^{210}Pb$	P	Qt	Urrego et al. (2013)

(continued)

Table 2.4 (continued)

N	Site/area	Map	Country	ArXiv	Age	Dating method	Proxies	Data	References
23	Anse Trabaud	AT	Martinique	Archaeol	L Holocene	^{14}C	A/St/Pt/Am	Qt	Hofman et al. (2021)
24	Pozo	Po	Colombia	Sed core	L Holocene	^{14}C	P/G/Pm	Qt	García et al. (2022)
25	Barkers	Ba	Cayman Is	Sed core	L Holocene	^{14}C	S	Ql	Woodroffe (1982)
26	Santa Marta	SM	Colombia	Sed core	L Holocene	^{14}C	P/Py	Qt	Cohen and Wiedemann (1973)
27	Morelos	Mo	Mexico	Sed core	L Holocene	^{14}C	P	Qt	Islebe et al. (2001)
28	Naranjo 1	RN	Guatemala	Sed core	L Holocene	^{14}C	P/Ch/Ph	Qt	Neff et al. (2006)
29	Yax Chen	YC	Mexico	Cave	L Holocene	^{14}C	S	Qt	Collins et al. (2015)
30	Grande	Gr	Puerto Rico	Sed core	L Holocene	^{14}C	P/Ch/S/D/F/G	Qt	Lane et al. (2013)
31	Lagartos	La	Mexico	Sed core	L Holocene	^{14}C	P	Qt	Carrillo-Bastos et al. (2013)
32	Petenes	Pe	Mexico	Sed core	L Holocene	^{14}C	P/G/Py/S	Qt	Gutiérrez-Ayala et al. (2012)
33	SC01	SC	Cuba	Sed core	L Holocene	^{14}C /^{135}Cs	P/Ch/F/S	Qt	Peros et al. (2015)
34	Changuinola	Cg	Panama	Sed core	L Holocene	^{14}C	P	Qt	Phillips et al. (1997)
35	Sipacate 1	St	Guatemala	Sed core	L Holocene	^{14}C	P/Ch/Ph	Qt	Neff et al. (2006)
36	Altona	Al	Virgin Is	Sed core	L Holocene	^{14}C	P/S/G	Qt	Jessen et al. (2008)
37	Tribugá	Tr	Colombia	Sed core	M Holocene	^{14}C	P/S	Qt	Jaramillo and Bayona (2000)
38	Palmar	Pa	Mexico	Sed core	M Holocene	^{14}C	P	Qt	Torrescano and Islebe (2006)
39	Turneffe	Tu	Belize	Sed core	M Holocene	^{14}C	P/G	Qt	Wooler et al. (2009)
40	Galeta	Ga	Panama	Sed core	M Holocene	^{14}C	P/G	Qt	Castañeda-Posadas et al. (2022)
41	Celestum	Ce	Mexico	Sed core	M Holocene	^{14}C /^{210}Pb	F/G	Qt	Hardage et al. (2022)

42	Naranjo 2	RN	Guatemala	Sed core	M Holocene	^{14}C	P/Ch/Ph	Qt	Neff et al. (2006)
43	Bluefields	Bl	Nicaragua	Sed core	M Holocene	^{14}C	P	Qt	McCloskey and Liu (2012)
44	Hondo	Ho	Mexico	Sed core	M Holocene	^{14}C	P/G	Qt	Aragón-Moreno et al. (2018)
45	Sipacate 2	St	Guatemala	Sed core	M Holocene	^{14}C	P/Ch/Ph	Qt	Neff et al. (2006)
46	Flamenco	Fl	Puerto Rico	Sed core	M Holocene	^{14}C	G	Qt	Cohen et al. (2016)
47	Tonalá	To	Mexico	Sed core	M Holocene	^{14}C	P/Dt/G/Py	Qt	Bocanegra-Ramírez et al. (2019)
48	Sipacate 3	St	Guatemala	Sed core	M Holocene	^{14}C	P/Ch/Ph	Qt	Neff et al. (2006)
49	Nariva	Na	Trinidad/ Tobago	Sed core	M Holocene	^{14}C	P	Qt	Ramcharan (2004)
50	Tigre	Ti	Venezuela	Sed core	M Holocene	^{14}C	P/Ch/Py/G	Qt	Montoya et al. (2019)
51	Encrucijada	En	Mexico	Sed core	M Holocene	^{14}C	P/Ch/F	Qt	Joo-Chang et al. (2015)
52	Nicoya	Ni	Costa Rica	Sed core	M Holocene	^{14}C	P	Qt	Horn (1985)
53	Navío Quebrado	NQ	Colombia	Sed core	M Holocene	^{137}Cs	P	Qt	Urrego et al. (2013)
54	Sibun	Si	Belize	Sed core	M Holocene	^{14}C	P/G	Qt	Monacci et al. (2011)
55	Manchón	Mh	Guatemala	Sed core	M Holocene	^{14}C	P/Ch/Ph	Qt	Neff et al. (2006)
56	Leche	Le	Cuba	Sed core	M Holocene	^{14}C	P/ Pm/F/O/G	Qt	Peros et al. (2007)
57	Aktun Ha	AH	Mexico	Sed core	M Holocene	^{14}C	P/G	Qt	Gabriel et al. (2009)
58	Maracas	Mr	Trinidad/ Tobago	Sed core	M Holocene	^{14}C	P/G	Qt	Ramcharan and McAndrews (2006)
59	Playa Medina	Me	Venezuela	Sed core	M Holocene	^{14}C	P	Qt	Rull et al. (1999)
60	Torani	Tn	Guyana	Outcrop	M Holocene	^{14}C	P	Qt	Van der Hammen (1963)
61	Oreala	Or	Guyana	Outcrop	M Holocene	^{14}C	P	Qt	Van der Hammen (1963)
62	Kwakwani	Kw	Guyana	Outcrop	M Holocene	^{14}C	P	Qt	Van der Hammen (1963)

(continued)

Table 2.4 (continued)

N	Site/area	Map	Country	ArXiv	Age	Dating method	Proxies	Data	References
63	Tzib	Tz	Mexico	Sed core	M Holocene	^{14}C	P/G	Qt	Carrillo-Bastos et al. (2013)
64	Mullins-Sapodilla	Mu	Belize	Sed core	M Holocene	^{14}C	S/G	Ql	Adomat and Gischler (2015)
65	Saladilla	Sa	Dominican Rep	Sed core	M Holocene	^{14}C	P/Ch/Dt	Qt	Caffrey et al. (2015)
66	Spanish Lookout	SL	Belize	Sed core	M Holocene	^{14}C	P/S/G	Qt	Monacci et al. (2009)
67	Twin Cays	TC	Belize	Sed core	M Holocene	^{14}C	S	Ql	MacIntyre et al. (2004)
68	Gatun	Gt	Panama	Sed core	E Holocene	^{14}C	P	Qt	Bartlett and Barghoorn (1973)
69	Cariaco	CB	Venezuela	Sed core	L Pleistocene	Stratigraphy	P	Qt	González and Dupont (2009)
70	San Pedro	SP	Mexico	Living	L Pleistocene	R/P	Mp	Qt	Aburto-Oropeza et al. (2021)
71	Ogle Bridge	OB	Guyana	Borehole	L Pleistocene	^{14}C	P/Ch/S	Qt	Van der Hammen (1963)
72	Port Morant	PM	Jamaica	Outcrop	L Pleistocene	ESR	S/Am	Ql	Mitchell et al. (2001)
73	Columbus	Cm	Trinidad/Tobago	Well/Outcrop	Pliocene	Biostratigraphy	P	Ql	Lamy (1986)
74	Orinoco	Or	Venezuela	Well/Outcrop	Pliocene	Biostratigraphy	P	Ql	Pocknall et al. (2001)
75	Talamanca	Tl	Costa Rica	Outcrop	Pliocene	Strat/Biostrat	P	Qt	Graham and Dilcher (1998)
76	Montañita	Mt	Venezuela	Well	Pliocene	Biostratigraphy	P	Qt	Lorente (1986)
77	Jocotán	Jc	Guatemala	Outcrop	Pliocene	Strat/Biostrat	P	Qt	Graham (1998)
78	AMF-1X	AF	Venezuela	Well	Mio-Pliocene	Biostratigraphy	P	Qt	Lorente (1986)
79	Mirebalais	Ms	Haiti	Outcrop	Mio-Pliocene	Strat/Biostrat	P	Qt	Graham (1990a)
80	Barinas	Br	Venezuela	Outcrop	Mio-Pliocene	Biostrat/Fission-track	P/S/Py	Ql	Bermúdez et al. (2017)

81	Gatun	Gt	Panama	Outcrop	Mio-Pliocene	Strat/Biostrat	P	Qt	Graham (1990b, 1991)
82	Urumaco	Ur	Venezuela	Well/ Outcrop	Mio-Pliocene	Biostratigraphy	P	Qt	Lorente (1986); Hambalek et al. (1994)
83	Tig-59X	T5	Venezuela	Well	L Mio-Pliocene	Biostratigraphy	P	Qt	Lorente (1986)
84	AMM-1X	A1	Venezuela	Well	L Miocene	Biostratigraphy	P	Qt	Lorente (1986)
85	Zamuro-1X	Za	Venezuela	Well	L Miocene	Biostratigraphy	P	Qt	Lorente (1986)
86	Guanipa-1X	Gu	Venezuela	Well	L Miocene	Biostratigraphy	P	Qt	Lorente (1986)
87	Hervidero-1X	Hv	Venezuela	Well	L Miocene	Biostratigraphy	P	Qt	Lorente (1986)
88	Veracruz	Ve	Mexico	Outcrop	L Miocene	Biostratigraphy	P	Qt	Graham (1976)
89	VLC-737	VC	Venezuela	Well	M/L Miocene	Biostratigraphy	P	Qt	Lorente (1986)
90	Cucaracha	Cc	Panama	Outcrop	M Miocene	Strat/Biostrat	P	Qt	Graham (1988b)
91	Yopal	Yp	Colombia	Outcrop	M Miocene	Biostratigraphy	P	Ql	Dueñas and Van der Hammen (2007)
92	TJ	TJ	Venezuela	Well	E Miocene	Biostratigraphy	P	Qt	Rull (2001)
93	Sheroli	Sh	Costa Rica	Outcrop	E Miocene	Strat/Biostrat	P	Qt	Graham (1987)
94	Gold Hill	GH	Panama	Outcrop	E Miocene	Strat/Biostrat	P	Qt	Graham (1988a)
95	SZZ-119X	SZ	Venezuela	Well	E Miocene	Biostratigraphy	P	Qt	Lorente (1986)
96	SCZ-124X	SC	Venezuela	Well	E Miocene	Biostratigraphy	P	Qt	Lorente (1986)
97	IZZ-100X	IZ	Venezuela	Well	E Miocene	Biostratigraphy	P	Qt	Lorente (1986)
98	Cascadas	Cd	Panama	Outcrop	E Miocene	Stat/Biostrat	P	Qt	Graham (1989)
99	B-188	B1	Venezuela	Well	E Miocene	Biostratigraphy	P	Qt	Lorente (1986)
100	Saltarín	Sn	Colombia	Well	E/M Miocene	Biostratigraphy	S/P	Qt	Jaramillo et al. (2017)
101	NZZ-154/160	NZ	Venezuela	Well	E/M Miocene	Biostratigraphy	P	Qt	Lorente (1986)
102	Los Pobres-1	LP	Venezuela	Well	E/M Miocene	Biostratigraphy	P	Qt	Lorente (1986)
103	Pariaguán-1X	Pg	Venezuela	Well	E/M Miocene	Biostratigraphy	P	Qt	Lorente (1986)
104	Panchita-1X	Pa	Venezuela	Well	E/M Miocene	Biostratigraphy	P	Qt	Lorente (1986)

(continued)

Table 2.4 (continued)

N	Site/area	Map	Country	ArXiv	Age	Dating method	Proxies	Data	References
105	SDZ-117X	SD	Venezuela	Well	E/M Miocene	Biostratigraphy	P	Qt	Lorente (1986)
106	LSJ-3310	LJ	Venezuela	Well	E/M Miocene	Biostratigraphy	P	Qt	Lorente (1986)
107	LS-992B	LS	Venezuela	Well	E/M Miocene	Biostratigraphy	P	Qt	Lorente (1986)
108	LL-370	LL	Venezuela	Well	E/M Miocene	Biostratigraphy	P	Qt	Lorente (1986)
109	AMN-2X	AN	Venezuela	Well	Miocene	Biostratigraphy	P	Qt	Lorente (1986)
110	Buena Vista	BV	Venezuela	Outcrop	Miocene	Biostratigraphy	P	Qt	Lorente (1986)
111	CO-85	CO	Trinidad/ Tobago	Well	Miocene	Biostratigraphy	P	Sq	Germeraad et al. (1968)
112	SLA	SL	Venezuela	Well	Miocene	Biostratigraphy	P	Qt	Rull (1992)
113	Culebra-1X	Cb	Venezuela	Well	Miocene	Biostratigraphy	P	Qt	Lorente (1986)
114	Catatumbo-1X	Ct	Venezuela	Well	Miocene	Biostratigraphy	P	Qt	Lorente (1986)
115	OG-1X	OG	Venezuela	Well	Miocene	Biostratigraphy	P	Qt	Lorente (1986)
116	Chafurray	Fy	Colombia	Well	Oligo-Miocene	Biostratigraphy	P	Sq	Germeraad et al. (1968)
117	Falcón	Fc	Venezuela	Well	Oligo-Miocene	Biostratigraphy	P	Qt	Rull and Poumot (1997)
118	Maturín	Mn	Venezuela	Well	Oligo-Miocene	Biostratigraphy	P	Qt	Helenes and Cabrera (2002)
119	Chama	Ch	Venezuela	Outcrop	Oligo-Miocene	Biostratigraphy	P	Qt	Rull (1997a)
120	Shelter Belt	SB	Guyana	Well	Oligo-Miocene	Biostratigraphy	P	Qt	Van der Hammen and Wijmstra (1964)
121	Simojovel	Sj	Mexico	Outcrop	Oligo- Miocene	Biostratigraphy	P	Qt	Graham (1999)
122	Planeta Rica	Pn	Colombia	Well	L Oligocene-E Miocene	Biostratigraphy	P	Ql	Dueñas (1980)
123	Pozón-SX	Pz	Venezuela	Well	Oligo-Miocene	Biostratigraphy	P	Qt	Lorente (1986)
124	Tig-141X	Tg	Venezuela	Well	Oligo-Miocene	Biostratigraphy	P	Qt	Lorente (1986)
125	Friata-1X	Fr	Venezuela	Well	Oligo-Miocene	Biostratigraphy	P	Qt	Lorente (1986)
126	Carbón	Ca	Venezuela	Outcrop	Oligo-Miocene	Biostratigraphy	P	Qt	Lorente (1986)
127	Lares	Lr	Puerto Rico	Outcrop	Oligocene	Strat/Biostrat	P	Qt	Graham and Jarzen (1969)

128	Concentración	Cn	Colombia	Outcrop	L Eocene-E Oligocene	Biostratigraphy	P	Qt	Ochoa et al. (2012)
129	COT-1X	Cf	Venezuela	Well	L Eocene-Oligocene	Biostratigraphy	P	Qt	Rull (2003)
130	Bogotá	Bg	Colombia	Outcrop	L Eocene	Biostratigraphy	P	Qt	Ochoa et al. (2012)
131	Delicias	De	Venezuela	Outcrop	Eocene-Oligocene	Strat/Biostrat	P	Ql	Colmenares and Teran (1993)
132	Alcalde Díaz	AD	Panama	Outcrop	M/L Eocene	Strat/Biostrat	P	Qt	Graham (1985)
133	Tarra	Ta	Venezuela	Well	M/L Eocene	Biostratigraphy	P	Qt	Rull (1997b)
134	Maraguán	Mg	Cuba	Outcrop	M Eocene	Strat/Biostrat	P	Qt	Graham et al. (2000)
135	VLC	VC	Venezuela	Well	M Eocene	Biostratigraphy	P	Qt	Rull (1998)
136	Uribe	Ub	Colombia	Outcrop	M Eocene	Strat/Biostrat	P	Qt	Jaramillo and Dilcher (2001)
137	Prevención	Pv	Venezuela	Well	M Eocene	Biostratigraphy	P	Sq	Germeraad et al. (1968)
138	Chapelton	Cp	Jamaica	Well	M Eocene	Strat/Biostrat	P	Qt	Graham (1977)
139	Icotea	Ic	Venezuela	Well	M Eocene	Biostratigraphy	P	Sq	Germeraad et al. (1968)
140	Boscán	Bo	Venezuela	Well	M Eocene	Strat/Biostrat	P	Ql	Colmenares (1988)
141	Paz del Río	PR	Colombia	Outcrop	M Eocene	Biostratigraphy	P	Sq	Germeraad et al. (1968)
142	Sagu	Sg	Colombia	Outcrop	M Eocene	Biostratigraphy	D/F/P	Qt	De la Parra et al. (2021)
143	Tibú	Tb	Colombia	Outcrop	E/M Eocene	Strat/Biostrat	P	Qt	González Guzmán (1967)
144	Regadera	Rg	Colombia	Outcrop	E/M Eocene	Strat/Biostrat	P	Qt	Jaramillo and Dilcher (2001)
145	Maché	Mc	Venezuela	Outcrop	Eocene	Biostratigraphy	P	Qt	Rull (1999, 2000b)
146	Maracaibo	Mb	Venezuela	Well	Eocene	Biostratigraphy	P	Qt	Rull and Poumot (1997)
147	Nuevo Mundo	NM	Colombia	Outcrop	Eocene	Biostratigraphy	P	Ql	Rodríguez-Forero et al. (2012)
148	Cerrejón	Cj	Colombia	Well	Paleocene-Eocene	Biostratigraphy	P	Qt	Jaramillo et al. (2007)

(continued)

Table 2.4 (continued)

N	Site/area	Map	Country	ArXiv	Age	Dating method	Proxies	Data	References
149	Chama	Ch	Venezuela	Outcrop	Paleocene-Eocene	Strat/Biostrat	P	Qt	Pocknall and Erlich (2020)
150	Lobaterita	Lb	Venezuela	Outcrop	Paleocene-Eocene	Strat/Biostrat	P	Qt	Pocknall and Erlich (2020)
151	Piñalerita	Pñ	Colombia	Outcrop	Paleocene-Eocene	Biostratigraphy	P	Qt	Jaramillo and Dilcher (2001)
152	Rubio	Ru	Venezuela	Outcrop	L Cretaceous-M Eocene	Strat/Biostrat	P	Qt	Pocknall and Erlich (2020)
153	Lora	Lo	Venezuela	Well	L Cretaceous-M Eocene	Biostratigraphy	P	Sq	Germeraad et al. (1968)
154	San Luís	Ss	Colombia	Well	L Cretaceous	Biostratigraphy	P	Qt	Santos (2012)
155	Simacota	Sc	Colombia	Well	L Cretaceous	Strat/Biostrat	P	Ql	Montaño et al. (2016)
156	Buitrera	Bu	Colombia	Outcrop	L Cretaceous	Strat/Biostrat	D/P	Qt	Garzón et al. (2012)

Age: E, Early; M, Middle; L, Late. *Proxies*: A, archaeology (lithic/ceramic/shell/coral artifacts); Am, animal macroremains; Ch, charcoal; D, dinoflagellates; Dt, diatoms; F, foraminifers; G, geochemistry (organic matter, elemental analysis, stable isotopes); Mp, molecular phylogenetics and phylogeography; O, ostracods; P, pollen; Ph, phytoliths; Pm, plant macroremains; Py, physical properties (magnetic susceptibility, paleomagnetism, petrography/mineralogy); S, sedimentology/stratigraphy; St, starch; T, thecamoebians. *Data*: Ql, qualitative; Qt, quantitative; Sq, semiquantitative

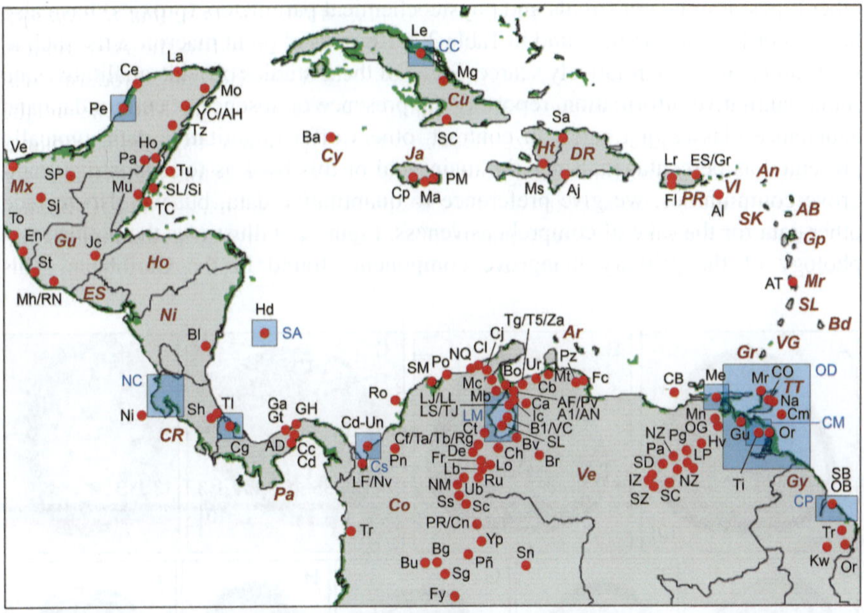

Fig. 2.8 Sketch map indicating the location of the sites used in this book (see Table 2.4 for abbreviations and details). Red dots indicate stratigraphic records (wells, outcrops, coring sites), and blue boxes indicate areas with modern-analog studies. The locations of many sites are approximate, due to the lack of coordinates in the original references. In these cases, the sites have been located with Google Earth using the descriptions provided in the original studies. *Countries (brown letters)*: Bz, Belize; Co, Colombia; CR, Costa Rica; Cu, Cuba; Cy, Cayman Islands (UK); DR, Dominican Republic; Gu, Guatemala; Gy, Guyana; Ht, Haiti; Ja, Jamaica; Mr., Martinique (France); Mx, Mexico; Ni, Nicaragua; Pa, Panama; PR, Puerto Rico; TT, Trinidad & Tobago; Ve, Venezuela; VI, Virgin Islands (USA/UK)

Venezuela, both of which have a long history of active hydrocarbon exploration. Within these areas, there is a wealth of data from the Eocene to Miocene, while records from the Paleocene and Oligocene are less common. The western and southwestern Caribbean region, including Colombia and Central America, is also well represented in the dataset. However, most of the studies in this region focus on Pleistocene and Holocene records, particularly in relation to factors such as climate and human activities that drive ecological changes. A similar predominance of Holocene studies is observed in the northern Caribbean coasts, primarily in the Greater Antilles. The eastern coasts, represented by the Lesser Antilles, have only one site from the French island of Martinique included in the dataset. It is worth noting that many of the studies lack specific geographical coordinates, often due to confidentiality rules. Consequently, the exact locations of these study sites are approximated, relying on published maps, detailed descriptions in the original references, and the use of Google Earth.

The CARMA dataset primarily consists of pollen studies, which are the most commonly used to reconstruct historical mangrove trends in the region. However, other types of indicator fossils and physicochemical parameters (proxies) have also been recorded and can be found in Table 2.4. Records of plant macrofossils, such as seeds and leaves, are relatively scarce. Some of these studies provide qualitative and semiquantitative information, reporting the presence or absence of certain data and abundance classes or ranges. In contrast, others offer quantitative data, typically presented as percentages. Since the main goal of this book is to reconstruct mangrove communities, we give preference to quantitative data, but we also include other data for the sake of comprehensiveness. Figure 2.9 illustrates the pollen morphology of the primary mangrove components found in the Caribbean. This

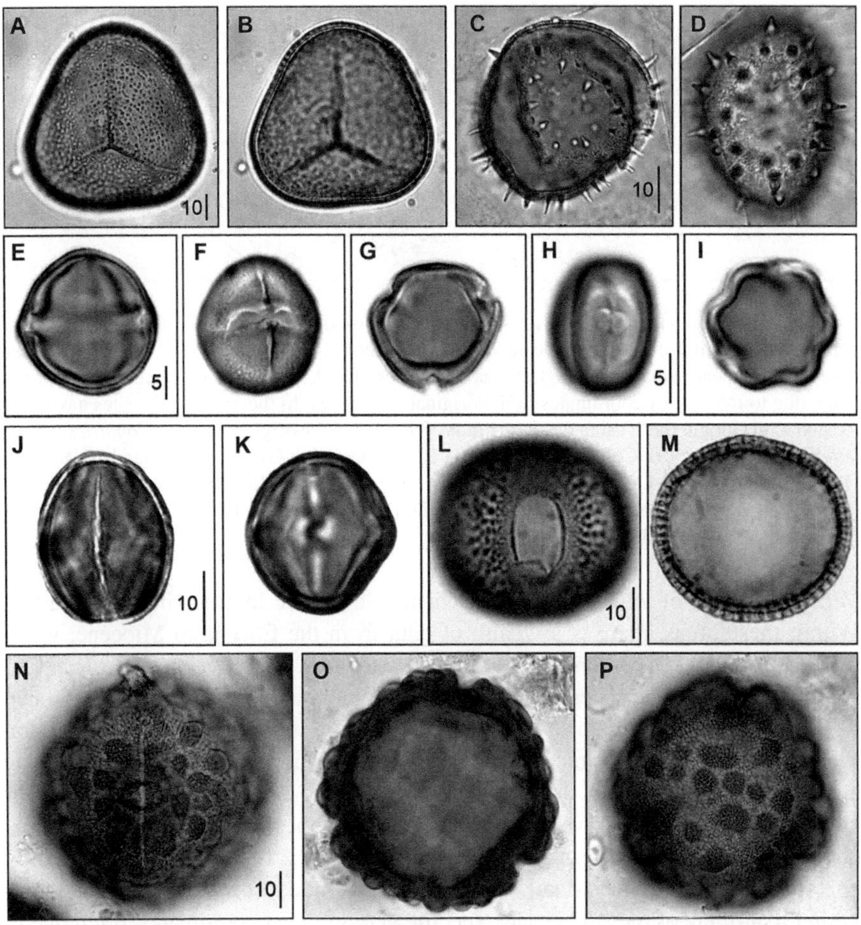

Fig. 2.9 Pollen/spores from the most representative Caribbean mangrove species (Table 2.2). (**a–b**) *Acrostichum aureum*; (**c–d**) *Nypa fruticans*; (**e–g**) *Rhizophora mangle*; (**h–i**) *Conocarpus erectus*; (**j–k**) *Laguncularia racemosa*; (**l–m**) *Avicennia germinans*; (**n–p**) *Pelliciera rhizophorae*. The palm *Nypa*, now restricted to the IWP region (Fig. 2.4), is included because it was part of Caribbean mangroves until the Eocene (Graham 1995; Gee 2001). Vertical bars are measurement scales in μm. After Rull (2022)

illustration is based on material obtained from living plants and pollen found in modern sedimentary deposits. It is important to note that pollen morphology is relatively uniform within each genus, and identifying species-level differences is often not feasible, except for a few ases. As a result, when discussing pollen, we typically refer to genera unless there is some discernible morphological differentiation at the species level.

The conventional practice in Quaternary paleoecology is to extend the generic names of existing mangrove components to the entire Quaternary period, assuming their presence over the last 2.6 Ma (Rull 2020). In instances where the presence of existing taxa cannot be confirmed in older sediments, artificial species (referred to as morphospecies) have been established based on pollen morphology. These morphospecies are then associated with existing genera, primarily determined by morphological similarities. Because pollen morphology is evolutionarily conservative (Erdtman 1986; Traverse 2007), it is traditionally believed that these morphospecies represent ancestors, likely at the generic level, of current species with similar ecological needs. Paleoecological studies often depend on the concept of niche constancy over time, particularly at the genus level, in long-lasting communities like mangroves (Wiens and Graham 2005; Hadly et al. 2009; Wiens et al. 2010). This approach, widely applied in plant evolution and particularly in the Neotropics (e.g., Flenley 1979; Morley 2000; Graham 2011), has been supported by recent molecular phylogenetic studies. These studies indicate that the primary Caribbean mangrove genera existed in the Paleogene, with their modern species predominantly emerging in the Neogene (Lo et al. 2014; Li et al. 2016; Duke 2017). Table 2.5 lists fossil representatives of the main existing mangrove genera, while the remaining true and associated mangrove genera (Table 2.2) lack known Cretaceous, Paleogene, or Neogene fossil counterparts and are found only in Quaternary and modern sediments. The

Table 2.5 Paleogene and Neogene fossil pollen representatives of extant mangrove genera from the Caribbean region

Genus	Fossil representative (morphospecies)	Range
Acrostichum	*Deltoidospora adriennis* (Potonié and Gelletich) Frederiksen	CPN
Avicennia	*Avicennia* type, *Retitricolporites* sp. Lorente	N
Crenea	*Verrutricoporites rotundisporus* Van der Hammen and Wijsmtra	N
Hibiscus	*Echiperiporites estelae* Germeraad, Hopping and Muller	N
Laguncularia	cf. *Laguncularia*	N
Nypa	*Spinizocolpites echinatus* Muller, *S. baculatus* Muller, *S. prominatus* (McIntyre) Stover and Evans	CP
Pachira	*Bombacacidites baculatus* Muller, Di Giacomo and Van Erve	N
Pelliciera	*Psilatricolporites crassus* Van der Hammen and Wijsmtra, *Lanagiopollis crassa* (Van der Hammen and Wijsmtra) Frederiksen	PN
Rhizophora	*Zonocostites ramonae* Germeraad, Hopping and Muller	PN

See Fig. 2.9 for true-mangrove elements and Fig. 2.10 for mangrove associates. Range: C, Cretaceous; P, Paleocene; N, Neogene. Based on Germeraad et al. (1968), Wijmstra (1968), Graham (1976), Muller (1981), Frederiksen (1985), Lorente (1986), Graham (2013), and Pocknall et al. (2022)

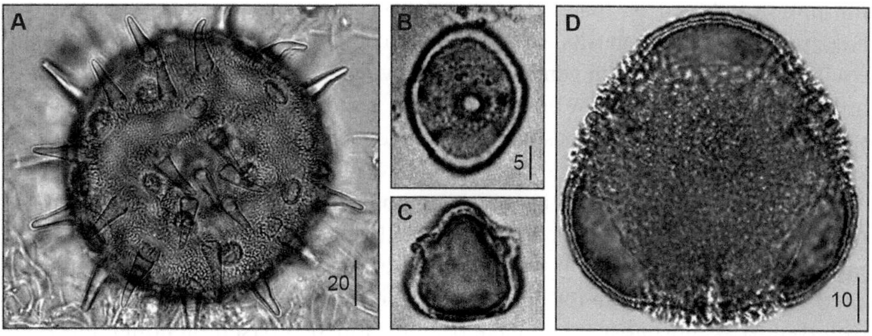

Fig. 2.10 Pollen from mangrove associates with fossil representatives (Table 2.5). (**a**) *Hibiscus tiliaceus* (Openly accessible at https://books.openedition.org/irdeditions/34279); (**b–c**) *Crenea patentinervis* (Modified from Lourteig 1986); (**d**) *Pachira aquatica* (Modified from Righetti et al. 2014). Vertical bars are measurement scales in μm

inclusion of the palm *Nypa fruticans* Wurmb, currently limited to the Indo-West Pacific region, is noteworthy. It was present in the Caribbean region until the Eocene (Graham 1995; Gee 2001), as explained in detail in the next chapters, where its biogeographical shifts and ecological requirements are also discussed. In some literature (e.g., Graham 2011), fossil pollen is referred to using the taxonomy of extant genera, rather than the corresponding fossil morphospecies. Here we maintain the terminology used in the cited literature to preserve the original data. In both cases, the equivalences provided in Table 2.5 prevent any potential confusion.

Appendix 1

Accompanying species—other than true and associated elements, which are listed in Table 2.2—identified in Neotropical mangroves (García-Fuentes et al. 2020). Nomenclature according to the International Plant Names Index (IPNI) (https://www.ipni.org/)

Family	Species	Plant type
Aizoaceae	*Sesuvium maritimum* Britton, Sterns & Poggenb.	Herb
	Sesuvium portulacastrum L.	Herb
Amaranthaceae	*Heterostachys ritteriana* Ung.-Sternb.	Shrub
	Salicornia bigelovii Torr.	Herb
	Salicornia ambigua Michx.	Herb
	Sarcocornia pacifica (Standl.) A.J.Scott	Herb
	Suaeda linearis (Elliot) Moq.	Herb
Annonaceae	*Annona glabra* L.	Tree
	Klarobelia anomala (R.E.Fr.) Chatrou	Tree

Family	Species	Plant type
Apocynaceae	*Blepharodon mucronatus* Decne.	Vine
	Clathrotropis macrocarpa Ducke	Tree
	Mesechites trifidus Müll.Arg.	Vine
	Sarcostemma clausum (Jacq.) Schult.	Vine
	Sarcostemma glaucum Kunth	Vine
Araceae	*Montrichardia linifera* Schott	Tree/Shrub
Arecaceae	*Bactris guineensis* (L.) H.E.Moore	Palm
	Coccothrinax litoralis León	Palm
	Roystonea borinquena O.F.Cook	Palm
	Roystonea regia O.F.Cook	Palm
	Sabal causiarum Becc.	Palm
	Sabal palmetto (Walter) Lodd. ex Schult. & Schult.f.	Palm
Asteraceae	*Ambrosia hispida* Pursch	Herb
	Baccharis halimifolia L.	Shrub
	Borrichia arborescens (L.) DC	Tree/Shrub
	Eclipta prostrata (L.) L.	Herb
	Iva cheiranthifolia Kunth	Herb
	Spilanthes urens Jacq.	Shrub/ Herb
	Wedelia fruticosa Jacq.	Herb
Bignoniaceae	*Bignonia hyacinthina* (Standl.) L.G.Lohmann	Vine
	Crescentia cujete L.	Tree
	Tabebuia angustata Britton	Tree
	Tabebuia rosea (Bertol.) DC.	Tree
Bromeliaceae	*Hohenbergia penduliflora* Mez	Epiphyte
	Tillandsia fasciculata Sw.	Epiphyte
	Tillandsia flexuosa Mez	Epiphyte
	Tillandsia recurvata (L.) L.	Epiphyte
	Tillandsia usneoides (L.) L.	Epiphyte
	Tillandsia variabilis Schltdl.	Epiphyte
Cactaceae	*Harrisia eriophora* (Pfeiff.) Britton	Cactus
	Opuntia dillenii Haw.	Cactus
	Pilosocereus royenii (L.) Byles & G.D.Rowley	Cactus
	Selenicereus grandiflorus (L.) Britton & Rose	Cactus
Capparaceae	*Crateva tapia* Burm.f.	Tree
	Quadrella odoratissima (Jacq.) Hutch.	Tree/Shrub
Combretaceae	*Bucida buceras* Vell.	Tree
	Bucida palustris Borhidi & O.Muñiz	Tree
	Terminalia catappa L.	Tree
Convolvulaceae	*Evolvulus convolvuloides* (Willd.) Stearn	Herb
	Ipomoea tiliacea Choisy	Vine
	Ipomoea triloba L.	Vine
Cucurbitaceae	*Cucumis melo* L.	Herb
	Melothria pendula L.	Herb

Family	Species	Plant type
Cyperaceae	*Cladium mariscus* R.Br.	Herb
	Cyperus alternifolius L.	Herb
	Eleocharis cellulosa Torr.	Herb
	Eleocharis interstincta R.Br.	Herb
	Eleocharis mutata (L.) Roem. & Schult.	Herb
	Fimbristylis cymosa R.Br.	Herb
	Fimbristylis spadicea (L.) Vahl.	Herb
	Kyllinga odorata Liebm.	Herb
	Schoenoplectus americanus (Pers.) Volkart ex Schniz & R. Keller	Herb
Fabaceae	*Acacia macracantha* Humb. & Bonpl. ex Willd.	Tree
	Albizia saman (Jacq.) F.Muell.	Tree
	Dalbergia berteroi Urb.	Shrub
	Dalbergia brownei (Jacq.) Urb.	Vine
	Desmanthus virgatus (L.) Willd.	Shrub/ Herb
	Erythrina fusca Lour.	Tree
	Gliricidia sepium Kunth	Tree
	Inga alba (Sw.) Willd	Tree
	Lonchocarpus pycnophyllus Urb.	Shrub
	Lonchocarpus sericeus (Poir.) Kunth ex DC.	Tree
	Machaerium lunatum Ducke	Tree/Shrub
	Mimosa pigra L.	Tree
	Pithecellobium mucronatum Britton in Shattuck	Tree/Shrub
	Prioria copaifera Griseb.	Tree
	Pterocarpus acapulcensis Rose	Tree
	Pterocarpus officinalis Jacq.	Tree
	Stahlia monosperma Urb.	Tree
Heliotropiaceae	*Heliotropium curassavicum* L.	Herb
Malvaceae	*Bastardia viscosa* (L.) Kunth	Shrub
	Cienfuegosia yucatanensis Millsp.	Herb
Moraceae	*Ficus bullenei* I.M.Johnst.	Tree
	Ficus dendrocida Kunth	Tree
	Maclura tinctoria (L.) D.Don ex Steud.	Tree
Nephrolepidaceae	*Nephrolepis multiflora* (Roxb.) F.M.Jarrett ex C.V.Morton	Fern
Onagraceae	*Ludwigia octovalvis* (Jacq.) P.H.Raven	Herb
Orchidaceae	*Broughtonia lindenii* (Lindl.) Dressler	Epiphyte
Phyllanthaceae	*Phyllanthus elsiae* Urb.	Tree

Family	Species	Plant type
Poaceae	*Chloris barbata* Sw.	Herb
	Chloris elata Desv.	Herb
	Distichlis spicata (L.) Greene	Herb
	Echinochloa polystachya (Kunth) Hitchc.	Herb
	Monanthochloe littoralis Engelm.	Herb
	Paspalum distachyon Willd. ex Döll.	Herb
	Phragmites australis (Cav.) Steud.	Reed
	Sporobolus pyramidatus (Lam.) Hitchc.	Herb
	Sporobolus virginicus (L.) Kunth	Herb
Polygonaceae	*Coccoloba uvifera* (L.) L.	Tree
	Persicaria acuminata (Kunth) M.Gómez	Herb
Pontederiaceae	*Eichhornia crassipes* (Mart.) Solms	Floating
Primulaceae	*Jacquinia keyensis* Mez	Shrub
Rubiaceae	*Casasia clusiifolia* Urb.	Tree/Shrub
	Erithalis fruticosa L.	Shrub
	Ernodea littoralis Sw.	Herb
	Morinda citrifolia L.	Tree
	Rachicallis americana Kuntze	Shrub
	Strumpfia maritima Jacq.	Shrub
Rutaceae	*Helietta plaeana* Tul.	Tree/Shrub
Salicaceae	*Casearia aculeata* Jacq.	Tree
Santalaceae	*Phoradendron quadrangulare* (Kunth) Griseb.	Shrub
Sapindaceae	*Paullinia fuscescens* Kunth	Vine
Solanaceae	*Lycium carolinianum* Walter	Shrub
	Lycium tweedianum Griseb.	Shrub
	Solanum jamaicense Mill.	Shrub
Surianaceae	*Suriana maritima* L.	Shrub
Typhaceae	*Typha domingensis* Pers.	Herb
Verbenaceae	*Lantana involucrata* L.	Shrub
Vitaceae	*Cissus trifoliata* Lour.	Vine
	Cissus verticillata (L.) Nicolson & C.E.Jarvis	Vine

References

Aburto-Oropeza O, Burelo-Campo CM, Ezcurra E et al (2021) Relict inland mangrove ecosystem reveals Last Interglacial sea levels. Proc Natl Acad Sci USA 118:e2024518118

Adomat F, Gischler E (2015) Sedimentary patterns and evolution of coastal environments during the Holocene in central Belize, Central America. J Coast Res 31:802–826

Aragón-Moreno AA, Islebe GA, Roy PD et al (2018) Climate forcings on vegetation of the southern Yucatán Peninsula (Mexico) during the middle to late Holocene. Palaeogeogr Palaeoclimatol Palaeoecol 495:214–226

Bartlett AS, Barghoorn ES (1973) Phytogeographic history of the Isthmus of Panama during the past 12,000 years (a history of vegetation, climate and sea-level change). In: Graham A (ed) Vegetation and vegetational history of Northern Latin America. Elsevier, Amsterdam, pp 203–299

Bermúdez MA, Hoorn C, Bernet M et al (2017) The detrital record of late-Miocene to Pliocene surface uplift and exhumation of the Venezuelan Andes in the Maracaibo and Barinas foreland basins. Basin Res 29:370–395

Biswas SR, Biswas PL, Limon SH et al (2018) Plant invasion in mangrove forests worldwide. Forest Ecol Manag 429:480–492

Blanco JF, Estrada EA, Ortiz LF et al (2012) Ecosystem-wide impacts of deforestation in mangroves: the Urabá Gulf (Colombian Caribbean) case study. ISRN Ecol 2012:958709

Bocanegra-Ramírez DM, Li H-C, Domínguez G et al (2019) Holocene climate change and sea level oscillations in the pacific coast of Mexico. Quat Int 528:100–108

Bunting P, Rosenqvist A, Hilarides L et al (2022) Global Mangrove Watch: updated 2010 mangrove forest extent (v2.5). Remote Sens 10:1669

Burgess C, Taylor M, Spencer N et al (2018) Estimating damages from climate-related natural disasters for the Caribbean at 1.5 °C and 2 °C global warming above preindustrial levels. Reg Environ Chang 18:2297–2312

Caffrey MA, Horn SP, Orvis KH et al (2015) Holocene environmental change at Laguna Saladilla, coastal north Hispaniola. Palaeogeogr Palaeoclimatol Palaeoecol 436:9–22

Campbell JD, Taylor MA, Stephenson TS et al (2011) Future climate of the Caribbean from a regional climate model. Int J Climatol 31:1866–1878

Carrillo-Bastos A, Islebe GA, Torrescano-Valle N (2013) 3800 years of quantitative precipitation reconstruction from the northwest Yucatan Peninsula. PLoS One 8:e84333

Castañeda-Posadas C, Correa-Metrio A, Escobar J et al (2022) Mid to late Holocene sea-level rise and precipitation variability recorded in the fringe mangroves of the Caribbean coast of Panama. Palaeogeogr Palaeoclimatol Palaeoecol 592:110918

Castaño A, Urrego L, Bernal G (2010) Dinámica del manglar en el complejo lagunar de Cispatá (Caribe colombiano) en los últimos 900 años. Rev Biol Trop 58:1347–1366

Church JA, White NJ (2006) A 20th century acceleration in global sea-level rise. Geophys Res Lett 33:94–97

Cohen A, Wiedemann HU (1973) Distribution and depositional history of some pre-lagoonal Holocene sediments in the Ciénaga Grande de Santa Marta, Colombia. Mitt Inst Colombo-Alemán Invest Cient 7:139–154

Cohen MCL, Lara RJ, Cuevas E et al (2016) Effects of sea-level rise and climatic changes on mangroves from southwestern littoral PR during the middle and late Holocene. Catena 143:187–200

Collins SV, Reinhardt EG, Werner CL et al (2015) Late Holocene mangrove development and onset of sedimentation in the Yax Chen cave system (Ox Bel Ha) Yucatan, Mexico: implications for using cave sediments as a sea-level indicator. Palaeogeogr Palaeoclimatol Palaeoecol 438:124–134

Colmenares OA (1988) A palynological study of samples from three wells of the Boscan Field, Venezuela. Rev Técnica INTEVEP 8:83–97

Colmenares OA, Teran L (1993) A biostratigraphic study of Paleogene sequences in southwestern Venezuela. Palynology 17:67–89

Dangremond EM, Feller IC, Sousa WP (2015) Environmental tolerances of rare and common mangroves along light and salinity gradients. Oecologia 179:1187–1198

Davidson DE (2007) Modern pollen spectra from mangrove ecosystems of the sabana-camagüey archipelago and Ciego de Ávila, Cuba. MS diss, Univ Toronto, Canada

De la Parra F, Pinzon D, Mantilla-Duran F et al (2021) Marine-lacustrine systems during the Eocene in northern South America – palynological evidence from Colombia. J S Am Earth Sci 108:103188

DeYoe H, Lonard RI, Judd FW et al (2020) Biological flora of the tropical and subtropical intertidal zone: literature review for *Rhizophora mangle* L. J Coast Res 36:857–884

Dueñas H (1980) Palynology of Oligocene-Miocene strata of borehole Q-E-22, Planeta Rica, northern Colombia. Rev Palaeobot Palynol 30:313–328

Dueñas H, Van der Hammen T (2007) Significado geológico y asociaciones palinológicas de las formaciones Diablo Inferior (Mioceno Tardío) y San Fernando Superior (Mioceno Medio), piedemonte cuenca de los Llanos Orientales, Colombia. Rev Acad Colomb Cienc 31:481–498

Duke NC (2017) Mangrove floristics and biogeography revisited: further deductions from biodiversity hot spots, ancestral discontinuities, and common evolutionary processes. In: Rivera-Monroy VH et al (eds) Mangrove ecosystems: a global biogeographic perspective. Springer, Berlin, pp 17–53

Duke NC (2020) A systematic revision of the vulnerable mangrove genus *Pelliciera* (Tetrameristaceae) in equatorial Ammerica. Blumea 65:107–120

Duke NC, Meyneccke J-O, Dittman S et al (2017) A world without mangroves? Science 317:41b–42b

Durán-Quesada AM, Sori R, Ordoñez P et al (2020) Climate perspectives in the Intra-Americas Seas. Atmos 11:959

Ellison AM, Farnsworth EJ, Merkt RE (1999) Origins of mangrove ecosystems and the marine biodiversity anomaly. Glob Ecol Biogeogr 8:95–115

Erdtman G (1986) Pollen morphology and plant taxonomy. Angiosperms. EJ Brill, Leiden

Escarraga-Paredes DS, Torrescano-Valle N, Islebe GA (2014) Análisis de la relación vegetación-lluvia de polen actual de las comunidades vegetales en el nororoeste de la Península de Yucatán, Mexico. Polibotánica 38:27–52

Flenley JR (1979) The equatorial rain forest: a geological history. Butterworths, London

Frederiksen NO (1985) Review of early Tertiary sporomorph paleoecology. Am Assoc Strat Palynol Contr Ser 19:1–92

Gabriel JJ, Reinhardt EG, Peros MC et al (2009) Palaeoenvironmental evolution of cenote Aktun Ha (Carwash) on the Yucatan Peninsula, Mexico and its response to Holocene sea-level rise. J Paleolimnol 42:199–213

García Y, Rangel J, Jaramillo A (2022) Environmental changes during the last 1800 years in the Neguanje mangrove, Tayrona National Natural Park, Colombian Caribbean. Rev Acad Colomb Cien Exact Fís Nat 46:90–107

García-Fuentes A, Lendínez-Barriga ML, Torres-Cordero JA et al (2020) A study on the mangrove formations of the Neotropical-Austroamerican Kingdom. Phytocoenologia 50:137–162

Garzón S, Warny S, Bart PJ (2012) A palynological and sequence-stratigraphic study of Santonian-Maastrichtian strata from the Upper Magdalena Valley basin in central Colombia. Palynology 36(Suppl 1):112–133

Gee CT (2001) The mangrove palm *Nypa* in the geologic past of the New World. Wetl Ecol Manag 9:181–194

Gentry AH (1982) Phytogeographic patterns as evidence for a Choco refuge. In: Prance GT (ed) Biological diversification in the tropics. Columbia University Press, New York, pp 112–136

Germeraad JH, Hopping CA, Muller J (1968) Palynology of Tertiary sediments from tropical areas. Rev Palaeobot Palynol 6:189–348

González C, Dupont LM (2009) Tropical salt marsh succession as sea-level indicator during Heinrich events. Quat Sci Rev 28:939–946

González Guzmán AE (1967) A Palynological Study of the Upper Los Cuervos and Mirador formations (Lower and Middle Eocene), Tibú Area, Colombia. EJ Brill, Amsterdam

González C, Urrego LE, Martínez JI et al (2010) Mangrove dynamics in the southwestern Caribbean since the 'Little Ice Age': a history of human and natural disturbances. The Holocene 20:849–861

Graham A (1976) Studies in Neotropical paleobotany. II. The Miocene communties of Veracruz. Mexico Ann Missouri Bot Gard 63:787–842

Graham A (1977) New records of *Pelliciera* (Theaceae/Pellicieraceae) in the Tertiary of the Caribbean. Biotropica 9:48–52

Graham A (1985) Studies in Neotropical paleobotany. IV. The Eocene communities of Panama. Ann Missouri Bot Gard 72:504–534

Graham A (1987) Miocene communities and paleoenviroments of southern Costa Rica. Am J Bot 74:1501–1518

Graham A (1988a) Studies in Neotropical paleobotany. V. The lower Miocene communities of Panama-the Culebra Formation. Ann Missouri Bot Gard 75:1440–1466

Graham A (1988b) Studies in Neotropical paleobotany. VI. The lower Miocene communities of Panama-the Cucaracha Formation. Ann Missouri Bot Gard 75:1467–1479
Graham A (1989) Studies in Neotropical paleobotany, VII. The lower Miocene communities of Panama-the La Boca Formation. Ann Missouri Bot Gard 76:50–66
Graham A (1990a) Late Tertiary microfossil flora from the Republic of Haiti. Am J Bot 77:911–926
Graham A (1990b) New angiosperm records from the Caribbean Tertiary. Am J Bot 77:897–910
Graham A (1991) Studies in Neotropical botany. X. The Pliocene communities of Panama-composition, numerical representations, and paleocommunity paleoenvironmental reconstructions. Ann Missouri Bot Gard 78:465–475
Graham A (1995) Diversification of Gulf/Caribbean mangrove communities through Cenozoic time. Biotropica 27:20–27
Graham A (1998) Studies in Neotropical botany. XI. Late Tertiary vegetation and environments of southeastern Guatemala: palynofloras from the Mio-Pliocene Padre Miguel Group and the Pliocene Herrería Formation. Am J Bot 85:1409–1425
Graham A (1999) Studies in Neotropical botany. XIII. An Oligo-Miocene palynoflora from Simojovel (Chiapas, Mexico). Am J Bot 86:17–31
Graham A (2011) A natural history of the new world. The ecology and evolution of plants in the Americas. University Chicago Press, Chicago
Graham SA (2013) Fossil records in the Lythraceae. Bot Rev 79:48–145
Graham A, Dilcher DL (1998) Studies in Neotropical botany. XII. A palynoflora from the Pliocene Rio Banano Formation of Costa Rica and the Neogene vegetation of Mesoamerica. Am J Bot 85:1426–1438
Graham A, Jarzen DM (1969) Studies on Neotropical paleobotany. I. The Oligocene communities of Puerto Rico. Ann Missouri Bot Gard 56:308–357
Graham A, Cozadd D, Areces-Mallea A et al (2000) Studies in Neotropical paleobotany. XIV. A palynoflora from the Middle Eocene Saramaguacán Formation of Cuba. Am J Bot 87:1526–1539
Gutiérrez-Ayala LV, Torrescano-Valle N, Islebe GA (2012) Reconstrucción paleoambiental del Holoceno tardío de la reserva Los Petenes. Rev Mex Cien Geol 29:749–763
Hadly EA, Spaeth PA, Li C (2009) Niche conservatism above the species level. Proc Natl Acad Sci USA 106:19707–19714
Hambalek N, Rull V, Di Giacomo E et al (1994) Evolución paleoecológica y paleoambiental de la secuencia del Neógeno en el Surco de Urumaco, Estado Falcón. Estudio palinológico y litológico. Bol Soc Venez Geól 1–2:7–19
Hardage K, Street J, Herrera-Silveira JA et al (2022) Late Holocene environmental change in Celestum Lagoon, Yucatan, Mexico. J Paleolimnol 67:131–162
Helenes J, Cabrera D (2002) Oligo-Miocene palynomorph assemblages from eastern Venezuela. Palynology 27:5–25
Herrera DA, Ault TR, Carrillo CM et al (2020) Dynamical characteristics of drought in the Caribbean from observations and simulations. J Clim 33:10773–10797
Hofman C-C (2002) Pollen distribution in sub-Recent sedimentary environments of the Orinoco Delta (Venezuela) – an actuo-palaeobotanical study. Rev Palaeobot Palynol 119:191–217
Hofman CL, Pagán-Jiménez JR, Fiekd MH et al (2021) Mangrove archives: unravelling human-environment interactions from deeply buried deposits at the site Anse Trabaud, Martinique, Lesser Antilles (1290-780 cal BP). Environ Archaeol. https://doi.org/10.1080/14614103.2021.1921676
Horn SP (1985) Preliminary pollen analysis of Quaternary sedimentsfrom Deep Sea Drilling Project site 565, western Costa Rica. Init Rep Deep Sea Drill Proj 84:533–547
IPCC (2019) Global Warming of 1.5 °C. Cambridge University Press, Cambridge
Islebe GA, Villanueva-Gutiérrez R, Sánchez O (2001) Relación lluvia de polen-vegetación en selvas de Quintana Roo. Bol Soc Bot Mexico 69:31–38
Jaramillo C, Bayona G (2000) Mangrove distribution during the Holocene in Tribugá Gulf, Colombia. Biotropica 32:14–22

Jaramillo C, Dilcher DL (2001) Middle Paleogene palynology of Central Colombia, South America: a study of pollen and spores from tropical latitudes. Palaeontographica B 258:87–213

Jaramillo C, Bayona G, Pardo-Trujillo A et al (2007) The palynology of the Cerrejón Formarion (Upper Paleocene) of northern Colombia. Palynology 31:153–189

Jaramillo C, Romero I, D'Apolito C et al (2017) Miocene flooding events of western Amazonia. Sci Adv 3:e1601693

Jessen CA, Pedersen JBT, Bartholdy J et al (2008) A late Holocene palaeoenvironmental record from Altona bay, St. Croix. US Virgin Islands Geogr Tidsskr 108:59–70

Jones PD, Harpham C, Harris I et al (2015) Long-term trends in precipitation and temperature across the Caribbean. Int J Climatol 36:3314–3333

Joo-Chang JC, Islebe GA, Torrescano-Valle N (2015) Mangrove history during middle- and late-Holocene in Pacific south-eastern Mexico. The Holocene 25:651–662

Khan NS, Vane CH, Engelhart SE et al (2019) The application of $\delta^{13}C$, TOC and C/N geochemistry of mangrove sediments to reconstruct Holocene paleoenvironments and relative sea levels, Puerto Rico. Mar Geol 415:105963

Kuyl OS, Muller J, Waterbolk HT (1955) The application of palynology to oil geology with reference to western Venezuela. Geol Mijnb 3:49–76

Lamy A (1986) Plio-Pleistocene palynology and visual kerogen studies, Trinidad, W.I., with emphasis on the Columbus Basin. Geol Soc Trinidad and Tobago, Geol Conf Trans, pp 114–127

Lane CS, Clark JJ, Knudsen A et al (2013) Late-Holocene paleoenvironmental history of bioluminiscent Laguna Grande, Puerto Rico. Palaeogeogr Palaeoclimatol Palaeoecol 369:99–113

LeBlanc AR, Kennedy LM, Liu K-B et al (2017) Linking hurricane landfalls, precipitation variability, fires, and vegetation response over the past millennium from analysis of coastal lagoon sediments, southwestern Dominican Republic. J Paleolimnol 58:135–150

Li X, Duke NC, Yang Y et al (2016) Re-evaluation of phylogenetic relationships among species of the mangrove genus *Avicennia* from Indo-West Pacific based on multilocus analyses. PLoS One 11:e0164453

Lo EYY, Duke NC, Sun M (2014) Phylogeographic pattern of *Rhizophora* (Rhizophoraceae) reveals the importance of both vicariance and long-distance oceanic dispersal to modern mangrove distribution. BMC Evol Biol 14:83

Lonard RI, Judd FW, Summy KR et al (2017) The biological flora of coastal dunes and wetlands: *Avicennia germinans* (L.) L. J Coast Res 33:191–207

Lonard RI, Judd FW, DeYoe H et al (2020a) Biology and ecology of the halophyte *Laguncularia racemosa* (L.) Gaertn. f.: a review. In: Grigore M-N (ed) Handbook of halophytes. Cham, Springer Nature, pp 1–6

Lonard RI, Judd FW, DeYoe H et al (2020b) Biology of the mangal halophyte *Conocarpus erectus* L.: a review. In: Grigore MN (ed) Handbook of halophytes. Cham, Springer Nature, pp 1819–1831

Lorente MA (1986) Palynology and palynofacies of the Upper Tertiary in Venezuela. Dissert Bot 99:1–222

Lourteig A (1986) Revisión del género *Crenea* Aublet (Litráceas). Caldasia 15:121–142

Lugo AE, Snedaker SC (1974) The ecology of mangroves. Ann Rev Ecol Syst 5:39–64

MacIntyre IG, Toscano MA, Lighty RG et al (2004) Holocene history of the mangrove islands of Twin Cays, Belize, Central America. Atoll Res Bull 510:1–16

Magaña V, Amador JA, Medina S (1999) The midsummer drought over Mexico and Central America. J Clim 12:1577–1588

Martínez C, Goddard L, Kushnir Y et al (2019) Seasonal climatology and dynamical mechanisms of rainfall in the Caribbean. Clim Dyn 53:825–846

McCloskey TA, Liu K-B (2012) A sedimentary-based history of hurricane strikes on the southern Caribbean coast of Nicaragua. Quat Res 78:454–464

Medina E, Cuevas E, Popp M et al (1990) Soil salinity, sunexposure, and growth of *Acrostichum aureum*, the mangrove fern. Bot Gaz 151:41–49

Medina E, Fernandez W, Barbosa F (2015) Element uptake accumulation and resorption in leaves of mangrove species with different mechanisms of salt regulation. Web Ecol 15:3–13

Mitchell SF, Pickerill RK, Stemann TA (2001) The Port Moran Formation (Upper Pleistocene, Jamaica): high resolution sedimentology and paleoenvironmental analysis of a mixed carbonate clastic lagoonal succession. Sedim Geol 144:291–306
Mona (Climate Studies Group) (2020) The State of the Caribbean Climate. Univ West Indies and Caribbean Development Bank, Kingston
Monacci NM, Meier-Grünhagen U, Finney BP et al (2009) Mangrove ecosystem changes during the Holocene at Spanish Lookout Cay, Belize. Palaeogeogr Palaeoclimatol Palaeoecol 280:37–46
Monacci NM, Meier-Grünhagen U, Finney BP et al (2011) Paleoecology of mangroves along the Sibun River, Belize. Quat Res 76:220–228
Montaño PC, Nova G, Bayona G et al (2016) Análisis de secuencias y procedencia en sucesiones sedimentarias de grano fino: un ejemplo de la Formación Umir y base de la Formación Lisama, en el sector de Simacota (Santander, Colombia). Bol Geol 38:51–72
Montoya E, Pedra-Méndez J, García-Falcó E et al (2019) Long-term vegetation dynamics of a tropical megadelta: mid-Holocene palaeoecology of the Orinoco Delta (NE Venezuela). Quat Sci Rev 221:105874
Morley RJ (2000) Origin and evolution of tropical rain forests. Wiley, Chichester
Muller J (1959) Palynology of recent Orinoco delta and shelf sediments. Micropaleontology 5:1–32
Muller J (1981) Fossil pollen records of extant angiosperms. Bot Rev 47:1–140
Muller J, Di Giacomo E, Van Erve AW (1987) A palynological zonation for the cretaceous, tertiary and quaternary of northern South America. Am Assoc Strat Palynol Contr Ser 19:7–76
Neff H, Pearsall DM, Jones JG et al (2006) Early Maya adaptive patterns: Mid-Late Holocene paleoenvironmental evidence from Pacific Guatemala. Latin Am Antiq 17:287–315
Ochoa D, Hoorn C, Jaramillo C et al (2012) The final phase of tropical lowland conditions in the axial zone of the Eastern Cordillera of Colombia: evidence from three palynological records. J S Am Earth Sci 39:157–169
Palanisamy H, Becker M, Meyssignac B et al (2012) Regional sea level change and variability in the Caribbean Sea since 1950. J Geodet Sci 2:125–133
Palmer SE, Burn MJ, Holmes J (2020) A multiproxy analysis of extreme wave deposits in a tropical coastal lagoon in Jamaica, West Indies. Nat Hazards 104:2531–2560
Palmieri S, Teodonio L, Siani A-M et al (2006) Tropical storm impact in Central America. Meteorol Appl 13:21–28
Peros MC, Reinhardt EG, Schwarcz HP et al (2007) High-resolution paleosalinity reconstruction from Laguna de la Leche, north coastal Cuba, using Sr, O, and C isotopes. Palaeogeogr Palaeoclimatol Palaeoecol 245:535–550
Peros M, Gregory B, Matos F et al (2015) Late-Holocene record of lagoon evolution, climate change, and hurricane activity from southeastern Cuba. The Holocene 25:1483–1497
Perrette M, Launderer F, Riva R et al (2013) Scaling approach to project regional sea level rise and its uncertainties. Earth Syst Dynam 4:11–29
Phillips S, Rouse GE, Bustin RM (1997) Vegetation zones and diagnostic pollen profiles of a coastal peat swamp, Bocas de Toro, Panama. Palaeogeogr Palaeoclimatol Palaeoecol 128:301–338
Pielke RA, Rubiera J, Landsea C et al (2003) Hurricane vulnerability in Latin America and The Caribbean: normalized damage and loss potentials. Nat Hazards Rev 4:101–114
Pocknall DT, Erlich RN (2020) Palynostratigraphy and lithostratigraphy of Upper Cretaceous and Paleogene outcrop sections, Mérida Andes (Maracaibo Basin), Western Venezuela. J S Am Earth Sci 104:102830
Pocknall DT, Wood LJ, Geen AF et al (2001) Integrated paleontological studies of Pliocene to Pleistocene deposits of the Orinoco Delta, Eastern Venezuela and Trinidad. In: Goodman DK, Clarke RT (eds) Proc IX Int Palynol Cong. Am Assoc Strat Palynol Found, Houston, Texas, pp 319–326
Pocknall DT, Clowes CD, Jarzen DM (2022) *Spinizonocolpites prominatus* (McIntyre) Stover & Evans: fossil *Nypa* pollen, taxonomy, morphology, global distribution, and paleoenvironmental significance. N Zealand J Geol Geophys 66:558–570

Rabinowitz D (1978) Early growth of mangrove seedlings in Panama and a hypothesis concernign the relationship of dispersal and zonation. J Biogeogr 5:113–133

Rahmstorf S (2007) A semi-empirical approach to projecting future sea level rise. Science 315:368–370

Ramcharan EK (2004) Mid-to-Late Holocene sea level influence on coastal wetland development in Trinidad. Quat Int 120:145–151

Ramcharan EK, McAndrews JH (2006) Holocene development of coastal wetland at Maracas Bay, Trinidad, West Indies. J Coast Res 22:581–586

Righetti VH, Ferreira CB, Gonçalves-Esteves V (2014) Pollen morphology of selected species of the subfamily Bombacoidae (Malvaceae *sensu lato*). Acta Bot Brasil 28:352–360

Rodgers JC, Horn SP (1996) Modern pollen spectra from Costa Rica. Palaeogeogr Palaeoclimatol, Palaeoecol 124:53–71

Rodríguez-Forero G, Oboh-Ikuenobe FE, Jaramillo C et al (2012) Palynology of the Eocene esmeraldas Formation, Middle Magdalena Valley Basin, Colombia. Palynology 36:96–111

Rull V (1992) Paleoecología y análisis secuencial de una sección deltaica terciaria de la cuenca de Maracaibo. Bol Soc Venez Geol 46:16–26

Rull V (1997a) Oligo-Miocene palynology of the Rio Chama sequence (western Venezuela), with comments on fossil algae as paleoenvironemntal indicators. Palynology 21:213–229

Rull V (1997b) Sequence analysis of western Venezuelan Cretaceous to Eocene sediments using palynology: chrono-paleoenvironmental and paleovegetational approaches. Palyology 21:79–90

Rull V (1998) Middle Eocene mangroves and vegetation changes in the Maracaibo basin, Venezuela. PALAIOS 13:287–296

Rull V (2001) A quantitative palynological record from the Early Miocene of western Venezuela, with emphasis on mangroves. Palynology 25:109–126

Rull V (2003) Contribution of quantitative ecological methods to the interpretation of stratigraphically homogeneous pre-Quaternary sequences: a palynological example from the Oligocene of Venezuela. Palynology 27:75–98

Rull V (2020) Quaternary ecology, evolution and biogeography. Elsevier/Academic Press, London

Rull V (2022) The Caribbean mangroves: an Eocene innovation with no Cretaceous precursors. Earth-Sci Rev 231:104070

Rull V (2023a) Taxon cycles in Neotropical mangroves. Plan Theory 12:244

Rull V (2023b) Rise and fall of Caribbean mangroves. Sci Total Environ 885:163851

Rull V (2023c) CARMA-F. Mendeley Data v3. https://doi.org/10.17632/zx8zvk3pw2.3

Rull V, Poumot C (1997) Eocene to Miocene palynocycles from western Venezuela and correlations with global eustatic cycles. Mem VII Cong Geol Venez II:343–349

Rull V, Vegas-Vilarrúbia T (1999) Surface palynology of a small coastal basin from Venezuela and potential paleoecological applications. Micropaleontology 45:365–393

Rull V, Vegas-Vilarrúbia T, Espinoza N (1999) Palynological record of an Early-Mid Holocene mangrove in eastern Venezuela. Implications for sea-level rise and disturbance history. J Coast Res 15:496–504

Santos CE (2012) Palynostratigraphy of the Umir Formation, Middle Magdalena Valley Basin (MMVB), Colombia. MS diss, Louisiana State Univ

Smith TJ (1992) Forest structure. In: Robertson AI, Alongi DM (eds) Tropical Mangrove ecosystems. Am Geophys Union, Washington, pp 101–136

Sousa WP, Kennedy PG, Mitchell BI et al (2007) Supply-side ecology in mangroves: do propagule dispersal and seedling establishment explain forest structure? Ecol Monogr 77:53–76

Stennett-Brown RK, Jones JJ et al (2017) Future Caribbean temperature and rainfall extremes from statistical downscaling. Int J Climatol 37:4828–4845

Taylor MA, Alfaro EJ (2005) Climate of Central America and the Caribbean. In: Oliver JE (ed) Encyclopedia of world climates. Springer, Dordrecht, pp 183–189

Taylor MA, Clarke LA, Centella A et al (2018) Future Caribbean climates in a world of rising temperatures: the 1.5 vs 2.0 dilemma. J Clim 31:2907–2926

Tomlinson PB (2016) The botany of Mangroves. Cambridge University Press, Cambridge

Torres RR, Tsimplis MN (2013) Sea-level trends and interannual variability in the Caribbean Sea. J Geophys Res 118:2934–2947
Torrescano N, Islebe GA (2006) Tropical forest and mangrove history from southeastern Mexico: a 5000 yr pollen record and implications for sea level rise. Veg Hist Archaeobotany 15:191–195
Traverse A (2007) Paleopalynology. Springer, Dordrecht
Tschudy RH (1969) Relationship of palynomorphs to sedimentation. In: Tschudy RH, Scott RA (eds) Aspects of palynology. Wiley Interscience, New York, pp 79–96
Urrego LE, Bernal G, Polanía J (2009) Comparison of pollen distribution patterns in surface sediments of a Colombian Caribbean mangrove with geomorphology and vegetation. Rev Palaeobot Palynol 156:358–375
Urrego LE, González C, Urán G et al (2010) Modern pollen rain in mangroves from San Andrés Island, Colombian Caribbean. Rev Palaeobot Palynol 162:168–182
Urrego LE, Correa-Metrio A, González C et al (2013) Contrasting responses of two Caribbean mangroves to sea-level rise in the Guajira Peninsula (Colombian Caribbean). Palaeogeogr Palaeoclimatol Palaeoecol 370:92–102
Urrego LE, Correa-Metrio A, González-Arango C (2018) Colombian Caribbean mangrove dynamics: anthropogenic and environmental drivers. Bol Soc Geol Mex 70:133–145
Urrego LE, Prado MA, Bernal G et al (2019) Mangrove responses to droughts since the little ice age in the Colombian Caribbean. Est Coast Shelf Res 230:196432
Van der Hammen TA (1963) palynological study on the Quaternary of British Guiana. Leidse Geol Mededel 29:125–168
Van der Hammen T, Wijmstra TA (1964) A palynological study on the tertiary and upper cretaceous of British Guiana. Leidse Geol Mededel 30:183–241
Webster PJ, Holland GJ, Curry JA et al (2005) Changes in tropical cyclone number, duration, and intensity in a warming environment. Science 309:1844–1846
Wiens JJ, Graham CH (2005) Niche conservatism: integrating evolution, ecology, and conservation biology. Annu Rev Ecol Evol Syst 36:519–539
Wiens JJ, Ackerly DD, Allen AP et al (2010) Niche conservatism as an emerging principle in ecology and conservation biology. Ecol Lett 13:1310–1324
Wijmstra TA (1968) The identity of *Psilatricolporites* and *Pelliciera*. Acta Bot Neerl 17:114–116
Wijmstra TA (1969) Palynology of the Alliance Well. Geol Mijn 48:125–133
Woodroffe CD (1982) Mangrove swamp stratigraphy and Holocene transgression, Grand Cayman Island, West Indies. Mar Geol 41:271–294
Wooler MJ, Behling H, Leon J et al (2009) Late Holocene hydrologic and vegetation changes at Turneffe Atoll, Belize, compared with records from mainland Central America and Mexico. PALAIOS 24:650–656
Yepes J, Poveda G, Mejía JF et al (2019) CHOCO-JEX a research experiment focused on the Chocó low-level jet over the far eastern Pacific and western Colombia. Bull Am Meteorol Soc 100:779–796

Eocene Origin

3

Abstract

The Caribbean mangroves, as part of Atlantic-Eastern Pacific (AEP) mangroves, are usually considered descendants of a purported Cretaceous pantropical mangrove belt that underwent regional differentiation after the tectonic closure of the Tethys Sea by the African barrier. This view, largely based on the scattered presence of *Nypa* and *Acrostichum* in a number of localities, changes if we use a quantitative community approach with emphasis on mangrove-forming trees. Such analysis reveals that there is no evidence for the occurrence of mangrove ecosystems during the Cretaceous and the Paleocene, and the first robust evidence for true mangrove communities appears in the Middle Eocene (~50–40 Ma). These first mangroves were dominated by the only mangrove-forming tree existing at that time, *Pelliciera*, and exhibited a characteristic sea–land zonation with *Nypa* and *Acrostichum* in the back-mangrove brackish swamps and some palms, including the salt-intolerant *Mauritia*, in the inland freshwater wetlands. *Pelliciera* first appeared along the southern Caribbean coasts in the Early Eocene and gained prominence in the Middle Eocene as it spread throughout the Caribbean region. This suggests that the initial mangroves in the Caribbean were ecological and evolutionary advancements that emerged de novo during the Eocene period, rather than being a result of regional evolutionary divergence from hypothetical Late Cretaceous Tethyan mangroves. The existence of these Tethyan mangroves is not substantiated by quantitative palynological records.

The distinct global distribution of mangroves, depicted in Fig. 2.4, has been ascribed to either dispersal or vicariance theories. Dispersal proponents posit that mangroves originated in the Indo-West Pacific (IWP) region and subsequently spread to the Atlantic-Eastern Pacific (AEP). Conversely, vicariance advocates argue that

V. Rull, *Origin and Evolution of Caribbean Mangroves*, Ecological Studies 252,
https://doi.org/10.1007/978-3-031-57612-6_3

mangrove species evolved in the Late Cretaceous near the continuous Tethys Sea (Fig. 3.1) and underwent diversification later. This diversification allegedly occurred when the closure of the Tethys Sea by the African barrier isolated the IWP and AEP regions. Supporting this perspective, Ellison et al. (1999) examined mangrove macrofossils (leaves, wood, flowers, fruits) and microfossils (pollen), along with mollusk fossils linked to mangrove ecosystems. They favored the vicariance theory, proposing that mangrove ecosystems originated at the Paleocene/Eocene boundary around the Tethys Sea, identifying three distinct regions of diversification: the Caribbean-Eastern Pacific, the Indian Ocean, and Southeast Asia. While the vicariance theory has gained broad acceptance, there are variations in proposed times of origin for pantropical mangrove communities and their separation into the AEP and IWP regions among different authors. For example, Plaziat et al. (2001) asserted that mangroves originated in the Paleocene, with the biogeographical split occurring at the Eocene/Oligocene boundary. Duke (2017) suggested a Late Cretaceous origin, followed by an Eocene/Oligocene split and further differentiation in the Late Miocene within the AEP and IWP regions. Srivastava and Prasad (2018) also placed the origins of mangroves in the Late Cretaceous, with the east-west split occurring at the Eocene/Oligocene boundary. According to these authors, mangrove

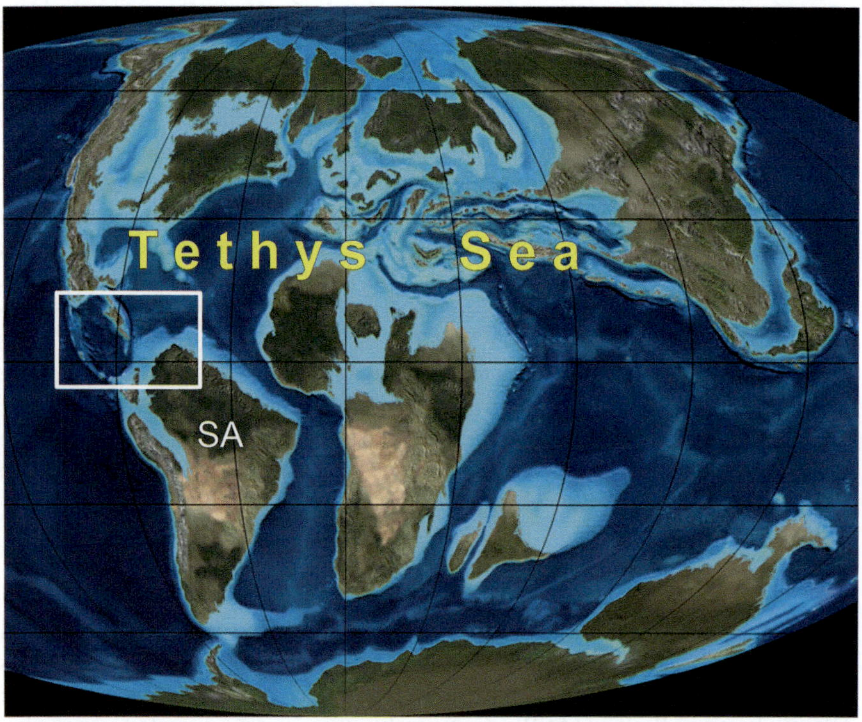

Fig. 3.1 Global continental configuration during the Late Cretaceous. The white box represents the approximate location of the present-day Caribbean region. SA, South America. Base map from https://en-academic.com/dic.nsf/enwiki/100955 (Mannion et al. 2014)

communities could have survived the Cretaceous/Paleocene mass extinction in tropical equatorial areas. Several studies, including Dodd and Afzal-Rafii (2002), Duke et al. (2002), Triest (2008), Duke (2017), and Van der Stocken et al. (2019), have provided detailed explanations of potential dispersal routes and barriers for mangroves since the Cretaceous. They have utilized molecular genetic evidence, simulations of oceanic paleocurrents, and considerations of global climate changes to support their arguments.

The phylogeographical patterns of *Rhizophora* have been employed as a template for comprehending the origin and diversification of mangroves across the planet. For example, Lo et al. (2014) utilized nuclear and chloroplastic DNA phylogenies to substantiate the concept of dispersal across the Tethys Sea, followed by a separation between the AEP and IWP lineages during the Eocene, and subsequent diversification within each biogeographical region during the Oligo-Miocene. However, alternative studies have proposed more recent timelines for the division of the AEP and IWP lineages. For instance, a recent investigation into phylogeography, utilizing chloroplastic DNA, proposed that the separation between the AEP and IWP lineages of *Rhizophora* occurred in the Middle Miocene, likely influenced by the complete closure of the Tethys seaway in the Oligo-Miocene and a global cooling period in the Miocene (Takayama et al. 2021). The application of molecular DNA-based phylogeography has proven valuable in elucidating the roles of dispersal and vicariance processes in the evolution of mangroves. Nonetheless, there is an ongoing necessity for further research, particularly in accurately dating these events through phylogenetic calibration. It is essential to approach this with caution to avoid circular reasoning, where phylogenetic trees are initially calibrated with well-documented geological events, and then the impact of these events on mangrove evolution is assessed.

In the Caribbean region, the main focus of this book, fossil pollen data suggested that the tropical mangrove communities underwent a gradual separate evolution from the pantropical flora of the Tethyan region during the transition from the Paleocene to the Eocene (Rull 1999, 2000). This divergence appeared to have been influenced by the global warming observed in the Early Eocene (Fig. 1.13). Jaramillo (2002) pointed out that this differentiation was likely a localized occurrence in northern South America, aligning with the concept of a distinct Eocene floral region referred to as "Neotropical" by Romero (1993). These findings would support the notion that the diversification of the AEP occurred around the Paleocene/Eocene boundary. The earliest fossil evidence of Neotropical mangrove taxa dates back to the Late Cretaceous (Srivastava and Prasad 2018), which was considered to be consistent with the idea of a pantropical mangrove distribution during those times.

However, examining the origins and development of mangroves requires a critical focus on identifying fossil evidence that can confidently reveal the existence of past mangroves as communities. The conventional method typically involves the use of botanical affinities between individual fossils (usually pollen and seeds) and extant mangrove taxa, usually at the genus level, as established, for example, by Muller (1981) and Frederiksen (1985). However, even if we assume the accuracy of these botanical resemblances, it does not guarantee a direct correlation between specific taxonomic and ecological characteristics. Consequently, drawing inferences about past mangrove ecosystems solely based on qualitative data indicating presence or absence of assumedly mangrove taxa is not justified, as emphasized by Plaziat et al. (2001). A more robust approach involves using fossil plant and animal assemblages indicative of mangrove communities as proxies for these ecosystems. Ideally, this approach should be quantitative, relying on the relative abundances of various components of mangrove ecosystems. However, achieving quantification is often challenging due to the distinct nature of fossils from different taxonomic groups and the varying conditions of their preservation.

A different method, the community approach, involves the reconstruction of ancient plant communities through the quantitative analysis of fossil pollen and spore assemblages. By analyzing the taxonomic composition, relative abundances, and spatial distribution of these fossils across the sea–land salinity gradient, researchers can compare them with the same patterns in contemporary mangrove ecosystems. This technique, known as the modern-analog approach, has been used in various studies across the Caribbean (e.g., Muller 1959; Rull 1998d; Hofmann 2002; Willard et al. 2001; Urrego et al. 2010; Polanía et al. 2015). Mangrove-forming trees play a crucial role in these reconstructions, and the identification of fossil assemblages containing these trees and the associated species in characteristic abundances similar to the present is considered a reliable indicator of past mangrove plant communities. This paleoecological evidence is typically found in coastal sediments, represented by rock formations characteristic of deltaic, shallow marine and other transitional land–sea environments. Quantitative pollen analysis, a widely used method in Quaternary paleoecology (Faegri et al. 1989; Moore et al. 1991), has significantly advanced the understanding of past ecological, biogeographical, and environmental dynamics (Birks and Birks 1981; Rull 2020). While the study of Quaternary mangrove ecosystems has benefited from this approach (e.g., Woodroffe and Grindrod 1991; Ellison 2008), the community-based approach has not been extensively applied to the reconstruction of Meso-Cenozoic mangrove ecosystems.

This chapter delves into pollen-based reconstructions of the southern Caribbean region, recognized as the birthplace of Neotropical mangroves. The aim is to identify the earliest documented mangrove assemblages using a community-based approach and to investigate the possible implications of these observations for the evolutionary trends and biogeographical patterns of Neotropical mangrove ecosystems. The initial part conducts a comprehensive examination of published quantitative palynological records spanning from the Late Cretaceous to the Eocene. The goal is to uncover hints indicating the presence of precursor mangrove communities, including their taxonomic composition and relative abundances. Following this, the gathered evidence is analyzed in detail to pinpoint the most ancient mangrove communities and their key characteristics. The ensuing discussion highlights the potential significance of these new findings in comprehending the evolution and distribution of mangroves within a global, pantropical context. Lastly, the chapter proposes some general recommendations for future research on the origins and development of mangrove ecosystems, in general.

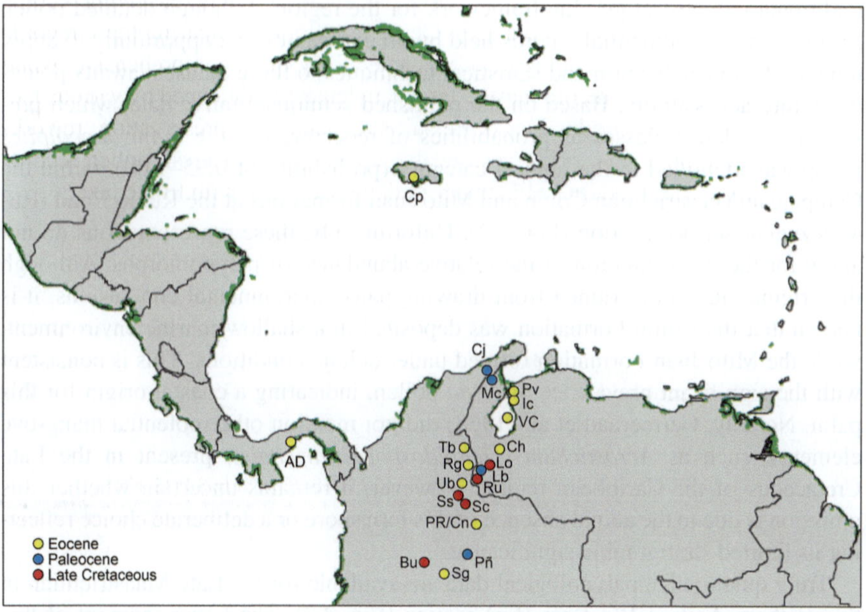

Fig. 3.2 Sketch map showing the location of the sites discussed in this chapter (abbreviations as in Table 2.4 and Fig. 2.8). The oldest age of each record is indicated

3.1 Quantitative Fossil Pollen Records

The earliest evidence of mangroves in the Caribbean area is found along the southern coasts, particularly in the northern parts of present-day Colombia and western Venezuela, as illustrated in Fig. 3.2. This evidence is based on the identification of fossil *Nypa* pollen (*Spinizonocolpites echinatus*/*S. prominatus*/*S. baculatus*) in Late Cretaceous (Maastrichtian) rock samples extracted from numerous oil wells in the region (Germeraad et al. 1968). In contrast, qualitative signs of mangroves in other Caribbean areas only began to appear in the Eocene (Frederiksen 1985; Graham 1995; Gee 2001). The following sections of this text will explore the quantitative data available on mangrove communities in the Caribbean, spanning from the Late Cretaceous to the Eocene. Many of these records were obtained through surveys conducted for the exploration and exploitation of natural resources, with palynology playing a crucial role in stratigraphy for both continental and coastal rock formations (Rull 2002; Jaramillo et al. 2005). The results will be presented first in the context of their original interpretations and subsequently discussed in a separate section.

3.1.1 Late Cretaceous

The initial documentation of Caribbean *Spinizonocolpites* was primarily qualitative or semiquantitative in nature, as the main focus of these studies was to establish a traditional palynostratigraphic framework for the region. Although detailed pollen counts exist in confidential records held by oil companies, the opportunity to apply modern data visualization and statistical techniques to these datasets awaits potential future accessibility. Based on the published semiquantitative data, which presented abundance classes as probabilities of reoccurrence, the *Spinizonocolpites* group was identified in the highest category (probability of 0.95–1.00) within the Campanian/Maastrichtian Colón and Mito Juan formations at the Rubio Road (Ru) Venezuelan surface section (Fig. 3.2). Unfortunately, these representations do not allow for the determination of the relative abundance of palynomorphs. Although the original authors refrained from drawing paleoenvironmental conclusions, it is known that the Colón Formation was deposited in a shallow marine environment, while the Mito Juan Formation formed under deltaic conditions. This is consistent with the significant prevalence of *Nypa* pollen, indicating a coastal origin for this palm. Notably, Germeraad et al. (1968) did not mention other potential mangrove elements, such as *Acrostichum* (*Deltoidospora adriennis*), present in the Late Cretaceous of the Caribbean region. However, it remains uncertain whether this omission is due to the actual absence of this fern spore or a deliberate choice reflecting its limited stratigraphic significance.

Truly quantitative palynological data are available for the Late Maastrichtian in the northern Andean Río Lora (Lo) section (Fig. 3.2). The lower segment of this area is characterized by the Late Maastrichtian Mito Juan Formation (Fig. 3.3), which was deposited in coastal marine environments (Pocknall and Erlich 2020). Within this formation, two distinct sections can be identified based on palynomorph content. In the lower portion, angiosperm pollen is the dominant feature, while dinoflagellates are moderately present, suggesting proximity to coastal settings (Fig. 3.4). Conversely, the upper portion exhibits higher levels of dinoflagellates and lower quantities of angiosperms, indicating more marine conditions, possibly in shallow marine environments. The prevalence of *Spinizonocolpites* is more notable in the lower segment and diminishes in the more marine-influenced part. Importantly, no other elements associated with mangroves were detected in the Mito Juan Formation of the Río Lora section.

A supplementary quantitative palynological investigation was carried out in the Umir Formation of Campanian/Maastrichtian age, recognized as equivalent to the Colón-Mito Juan formations (Dueñas 1989; Villamil 1999). This research was conducted in the San Luis (Ss) region within the Middle Magdalena Basin, where coal exploration necessitated the drilling of multiple cores. The examined composite section from these cores covered both the Middle and Upper Umir Formations, dating back to the Late Maastrichtian (Tchegliakova 1995). The primary focus of the study was the analysis of pollen, spores, and various other palynomorphs, including dinoflagellate cysts, acritarchs, and foraminiferal linings (Santos 2012). These palynomorphs were classified into five main groups: spores, palms, *Spinizonocolpites*, marine palynomorphs, and other angiosperms. Notably, fern spores and palms were the most prevalent and are believed to have been transported by rivers from inland environments. The *Spinizonocolpites* group, primarily characterized by *S. baculatus*, exhibited a relatively lower abundance (Fig. 3.5). The marine group, consisting

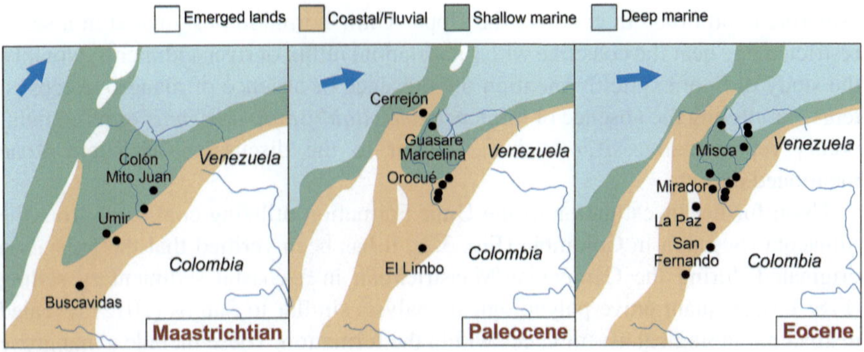

Fig. 3.3 Late Cretaceous to Eocene paleoenvironments in NW South America indicating the main formations. Localities (black dots) as in Fig. 3.2. The blue arrow indicates the movement of the Caribbean plate (Fig. 4.4). Modified from Rull (2022)

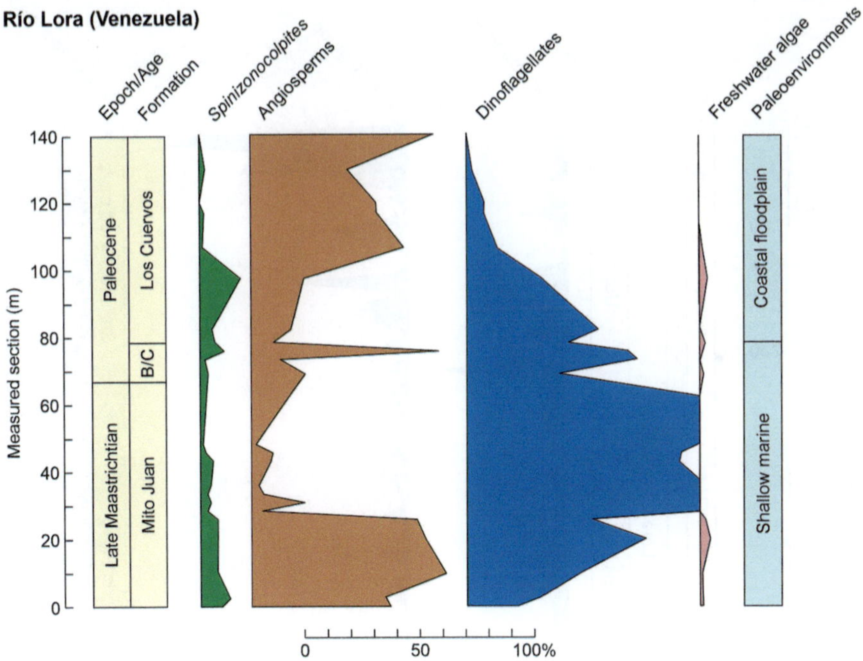

Fig. 3.4 Percentage diagram of selected palynomorphs from the Río Lora (Lo) section, western Venezuela (Fig. 3.2). B, Barco Fm; C, Catatumbo Fm. Redrawn and modified from Pocknall and Erlich (2020)

of dinoflagellate cysts, acritarchs, and foraminiferal linings, maintained consistent abundances throughout the section, except for a significant increase in the uppermost part. Based on these findings, the author concluded that the Middle Umir Formation was deposited in a lagoonal environment with coastal swamps and estuarine conditions. In contrast, the Upper Umir Formation originated in a semi-restricted bay near the coastline with a substantial influx of river sediments. Notably, the study did not explicitly mention the presence or absence of mangrove ecosystems, paralleling the absence of reference to *Deltoidospora adriennis* or its contemporary representative, *Acrostichum*, similar to the discussion of Rubio Road mentioned earlier.

Upon further investigation of the Umir Formation utilizing coal cores from the Simacota (Sc) area in Colombia (Fig. 3.2), it has been verified that the formation originated during the Campanian/Maastrichtian in a coastal sedimentary setting (Fig. 3.3). A quantitative palynological analysis similar to Santos (2012) revealed distinct palynological assemblages within the formation. These include a mangrove group identified by *Spinizonocolpites* species, a marine group comprising dinoflagellate cysts, acritarchs, foraminiferal linings and scolecodonts, and a freshwater

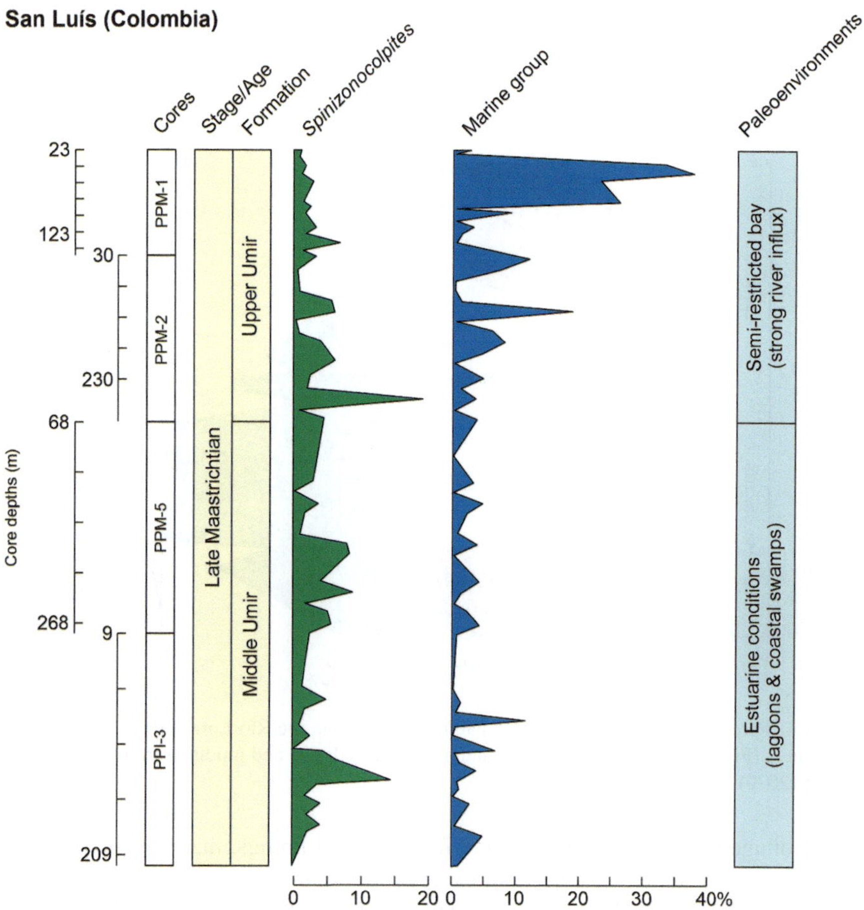

Fig. 3.5 Percentage diagram of the composite core section from San Luís (Ss), Magdalena Valley, Colombia (Fig. 3.2). Redrawn and modified from Santos (2012)

group containing the aquatic fern *Azolla* and the alga *Pediastrum* (Montaño et al. 2016). It was concluded that the Umir Formation was deposited in coastal swamps, transitioning to more continental environments in its upper layers. Notably, the *Spinizonocolpites* group was explicitly linked to mangrove communities. Unfortunately, no pollen diagrams were provided for comparative analysis. In the Upper Magdalena Valley, the Campanian/Maastrichtian Buscavidas Formation was palynologically examined in the Buitrera surface section (Bu), as illustrated in Fig. 3.2. The *Spinizonocolpites* group and *Deltoidospora adriennis* were absent, except for the presence of *Spinizonocolpites baculatus* in the uppermost layer (Garzón et al. 2012). According to these researchers, the absence of peridinioid cysts and the increase in estuarine (ceratioid and gymnodinoid cysts) and terrestrial (pollen and spores) palynomorphs indicated a progressive shallowing in the upper part of the Buscavidas Formation, which is consistent with the general paleoenvironmental trends in the study area (Fig. 3.3). Notably, there was no mention of the existence of mangrove communities in this region.

3.1.2 Paleocene

The aforementioned areas, Rubio Road and Río Lora, also host Paleocene sediments represented by the Barco, Catatumbo, and Los Cuervos formations, collectively constituting the Orocué Group. The Barco and Catatumbo formations were deposited in shallow marine environments near the coast, while the overlying Los Cuervos Formation was laid down in a floodplain to upper shoreface environment (Fig. 3.3) (Pocknall and Erlich 2020). In the Rubio Road section, the Catatumbo Formation exhibited the *Spinizonocolpites* group as the sole mangrove-related taxon, showing similar abundances to those in the Late Cretaceous section (Fig. 3.6). Within the Los Cuervos Formation, the assemblage was predominantly composed of non-mangrove species like the *Proxapertites* group, *Retidiporites magdalenensis*, and *Bombacacidites annae*, with *Spinizonocolpites* no longer present (Germeraad et al. 1968). In the Paleocene part of the Río Lora section, *Spinizonocolpites* reached its highest abundances during the transition from the Barco/Catatumbo to the Los Cuervos formations, signifying the shift from shallow marine to more coastal environments (Fig. 3.3). Scattered occurrences of *Spinizonocolpites* were also observed in two other Venezuelan sections, Río Lobaterita (Lb) and Río Chama (Ch) (Figs. 3.2 and 3.3). In these sections, dinoflagellate cysts were infrequent, and the Los Cuervos Formation exhibited more continental conditions (Pocknall and Erlich 2020).

The Riecito Maché (Mc) sequence (Fig. 3.2) is composed of sedimentary layers spanning from the Paleocene to the Middle Eocene. Within the Paleocene segment, the Guasare and Marcelina formations are prominent, overlaying Late Cretaceous sediments (Fig. 3.3). The Guasare Formation, recognized for its abundant coal deposits, originated in freshwater swamps and peat bogs characterized by fluctuating oxygen levels. It is hypothesized that angiosperms from inland regions were the primary contributors to peat formation in this environment (Escobar et al. 2016). Conversely, the Marcelina Formation is considered the lateral equivalent of the Los

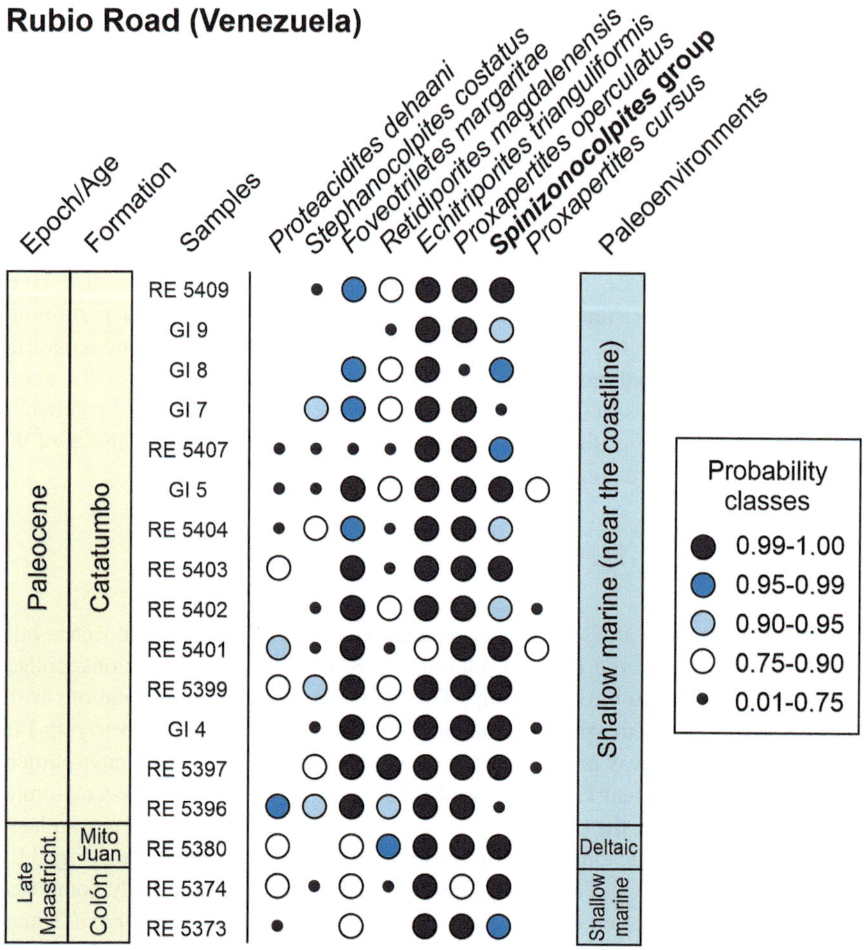

Fig. 3.6 Semiquantitative record of the main Late-Cretaceous to Paleocene palynomorphs from the Rubio Road (Ru) section, western Venezuela (Fig. 3.2). Redrawn and modified from Germeraad et al. (1968)

Cuervos Formation (Pocknall and Erlich 2020) and was deposited in a swampy setting emerging from a progradational deltaic sequence along the shelf margin of the Guasare Formation (González de Juana et al. 1980). A quantitative palynological analysis, utilizing statistically significant pollen counts (Rull 1987), identified three primary associations through cluster analysis (Rull 1999). The first assemblage, featuring *Proxapertites operculatus* and *Retidiporites magdalenensis*, is believed to originate from an undefined continental source. The second assemblage, comprising *Ctenolophonidites lisamae*, *Echitriporites trianguliformis* and two species of *Proxapertites*, suggests forested environments periodically flooded by freshwater. The third assemblage, marked by the presence of *Deltoidospora adriennis* and *Mauritiidites franciscoi* (related to the modern salinity-intolerant palm *Mauritia*), indicates a transitional zone between alluvial plains and coastal plains without tidal influence (Rull 1998a, b, c, d; Rull and Montoya 2014). The stratigraphic distribution of these assemblages implies a gradual shift from freshwater to coastal swamps (Fig. 3.7). Notably, the absence of *Spinizonocolpites* (*Nypa*) and the presence of *Deltoidospora adriennis* (*Acrostichum*) suggest that inland freshwater environments dominated and were unsuitable for *Nypa*. The discovery of *Spinizonocolpites* in Paleocene rocks within coastal and shallow marine sediments of the Rubio Road and Río Lora sediments (Figs. 3.6 and 3.4) is consistent with this interpretation.

In Colombia, during the Paleocene, the coastal plains bore a resemblance to those found in the Venezuelan Los Cuervos Formation of the Rubio Road section.

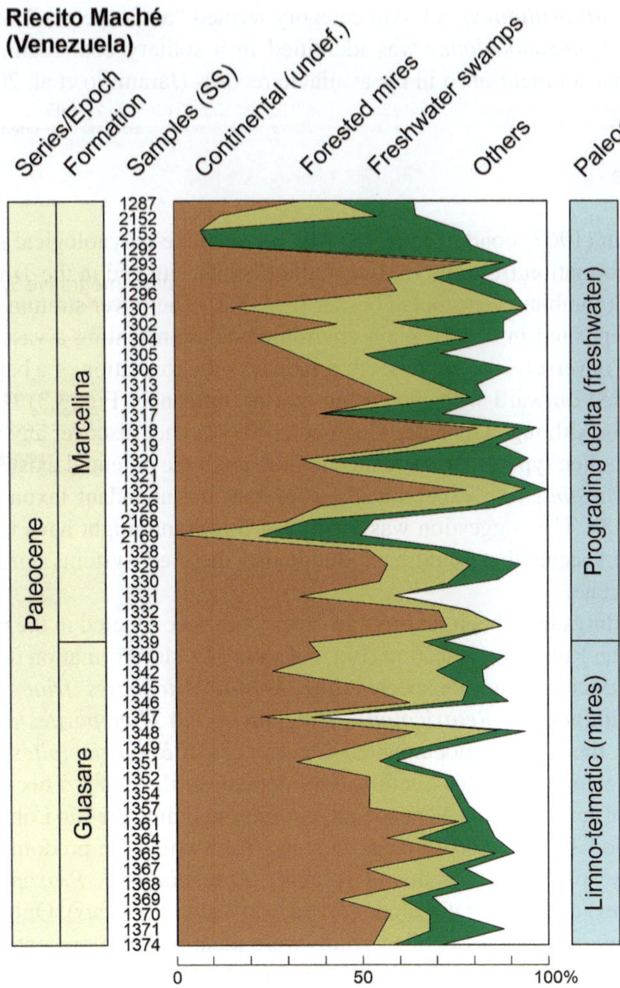

Fig. 3.7 Paleocene palynological assemblages defined in the Riecito Maché (Mc) section, Venezuela. Redrawn and modified from Rull (1999)

This similarity extended to the presence of common plant species, including *Proxapertites* spp., *Retidiporites magdalenensis*, and *Bombacacidites annae* (Jaramillo 2002). A specific example illustrating this connection is the Piñalerita (Pñ) section, where the El Limbo clays represent the Paleocene coastal plains (Fig. 3.2). Notably, *Spinizonocolpites* (*Nypa*) was scarce in this context, and *Deltoidospora* (*Acrostichum*) was not reported. It is important to note that available data only include pollen counts, lacking comprehensive pollen diagrams (Jaramillo and Dilcher 2001). The Paleocene presence is also evident in two wells at the Colombian El Cerrejón (Cj) coal mine (Figs. 3.2 and 3.3). In this case, the representation comes from the Cerrejón Formation, deposited in tidal flats, coastal lagoons, and flooded coastal plain environments. Pollen diagrams from this site reveal a prevalence of non-mangrove plant associations, such as *Proxapertites* spp., palm pollen (including *Mauritiidites*), a broad category termed "angiosperms," and fern spores. Notably, *Spinizonocolpites* was identified in a solitary coal seam, while *Deltoidospora* was not mentioned in the available records (Jaramillo et al. 2007).

3.1.3 Eocene

González Guzmán (1967) conducted pioneering quantitative palynological studies on the Mirador Formation from the Early-Middle Eocene, situated in the Tibú (Tb) region near the Colombian-Venezuelan border (Fig. 3.2). The lower stratum of this formation was deposited in coastal plain environments, representing a vast floodplain sequence. Conversely, the upper section indicates the formation of a bay-head delta that expanded outward with increasing marine influence (Fig. 3.3) (Ramon and Fajardo 2006). Although González Guzmán (1967) did not discover any identifiable mangrove pollen types, there was speculation about the potential existence of *Brevitricolpites variabilis*, an exceptionally common and abundant taxon in the Tibú area (Fig. 3.8). The suggestion was made that this taxon might have been an extinct mangrove species that could have dominated these ecosystems during the Early-Middle Eocene.

No similar findings have been reported in other studies conducted in the region. For example, in the Rubio Road section (Fig. 3.2), the Mirador Formation displays a prevalent presence of morphospecies like *Retibrevitricolpites triangulatus*, *Striatricolpites catatumbus*, *Retitricolpites irregularis*, and *Monoporites annulatus*. Interestingly, there are no documented occurrences of *Brevitricolpites* or any other mangrove elements in this section (Germeraad et al. 1968). This pattern remains consistent across various Venezuelan sections, including the Río Lobaterita, Río Lora, and Río Chama outcrops (Figs. 3.2 and 3.3), which are predominantly characterized by plants from fluvial plains (*Mauritiidites*, *Proxapertites*, *Cicatricosisporites*) and freshwater algae (*Pediastrum*, *Botryococcus*). Only a few occurrences of *Spinizonocolpites* pollen grains were identified as representative of mangrove elements (Pocknall and Erlich 2020). In the Colombian La Regadera (Rg) section (Figs. 3.2 and 3.3), a similar prevalence of inland assemblages was observed, with *Brevitricolpites* found in a single sample containing two pollen

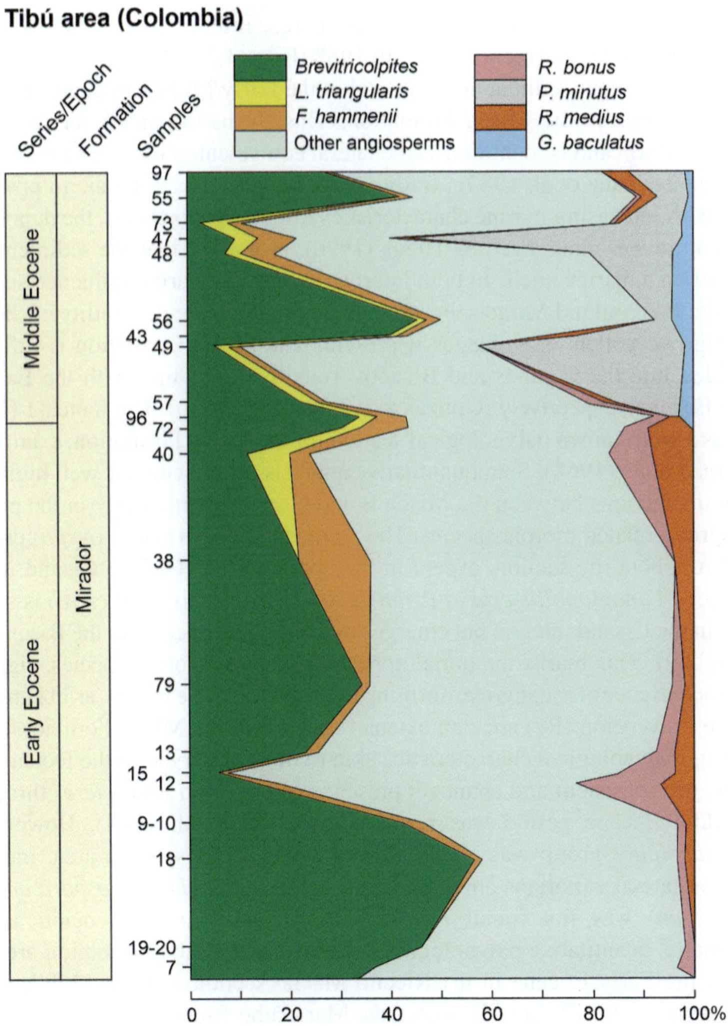

Fig. 3.8 Percentage pollen diagram from the Early–Middle Eocene of the Colombian Tibú (Tb) area (Fig. 3.2). Redrawn and modified from González Guzmán (1967). *F, Foveotriporites; G, Gemmamonocolpites; L, Longitricotomocolpites; P, Psilatricolpites; R, Retricolporites*

grains (Jaramillo and Dilcher 2001). In Colombia, the Early/Middle Eocene is also represented by the El Limbo/San Fernando and La Paz formations in the Piñalerita and Uribe (Ub) sections (Figs. 3.2 and 3.3). These sections exhibit a comparable pattern, although the Piñalerita section did register a low abundance of *Spinizonocolpites* (*Nypa*) species in the Middle Eocene. It is important to note that, again, pollen diagrams were not provided for these three sections, and only raw counts are available (Jaramillo and Dilcher 2001). The San Fernando Formation, located in the La Paz del Río (PR) section, shares similar palynological characteristics as reported by Germeraad et al. in 1968 (Figs. 3.2 and 3.3).

Another pertinent sedimentary unit within the Early/Middle Eocene strata in the investigated area is the Misoa Formation. This formation spans the Venezuelan Maracaibo Basin and is identified as the lateral equivalent of the Mirador Formation (González de Juana et al. 1980). Traditionally perceived as a deltaic to pro-deltaic deposit with increasing marine characteristics toward the northeast, the depositional Misoa nature was challenged by Higgs (1996), proposing that the sediments were deposited on a marine shelf. In both interpretations, the marine influence surpasses that of the more inland Mirador Formation, heightening the possibility of discovering mangrove pollen assemblages if present. The Misoa Formation is informally subdivided into the C-sands and B-sands, roughly coinciding with the Early and Middle Eocene, respectively (González de Juana et al. 1980). The Icotea-1 (Ic) well serves as a well-known palynological section for the Misoa Formation, examined by Germeraad et al. (1968). Semiquantitative analysis of the Icotea-1 well highlighted notable distinctions between the Misoa B and C sands, particularly in the presence of mangrove-related morphospecies. The *Spinizonocolpites* group (*Nypa*) predominates throughout the section, except in the uppermost part of the C-sand interval. Conversely, *Lanagiopollis crassa/Psilatricolporites crassus* (*Pelliciera*) is scarcely present in the C-sand interval but emerges as a dominant species in the B-sand interval (Fig. 3.9). This marks the initial appearance of this morphospecies, signifying the first occurrence of a mangrove-forming tree representative in the Caribbean region.

In the Prevención (Pv) area, an extensive record of the Misoa Formation exists, displaying palynological characteristics akin to those observed in the Icotea-1 well. Notably, the consistent and abundant presence of *L. crassa* (*Pelliciera*) throughout the Middle Eocene period was noted (Germeraad et al. 1968). However, the *Spinizonocolpites* group was almost non-existent in this specific area, indicating significant lateral variations among neighboring sections. *Deltoidospora adriennis* (*Acrostichum*) was not found in either the Icotea-1 or Prevención sections. Additionally, quantitative palynological studies of the Misoa Formation are accessible for the Early Eocene in the Riecito Maché section and the Middle Eocene (Lutetian) in a VLC (VC) well from Lake Maracaibo Block III. The Early Eocene part of the Riecito Maché section revealed three primary palynological assemblages (Rull 1999): inland marsh forests (marked by *Striatricolpites catatumbus*, *Retibrevitricolpites triangularis*, and *Retitricolpites irregularis*), undefined continental communities (*Proxapertites operculatus* and *Retitricolpites amapaensis*), and transitional fluvial/coastal marshes (*Deltoidospora adriennis*, *Mauritiidites franciscoi*, *Verrucatosporites speciosus*, and *Psilamonocolpites medius*). These assemblages exhibited varying trends without a clear stratigraphic pattern along the sequence, as illustrated in Fig. 3.10. Similar to the Paleocene interval in the same section, the coexistence of *Deltoidospora* (*Acrostichum*) and *Mauritiidites* (*Mauritia*) suggested that the coastal marshes and swamps in this sequence represented transitional environments between fluvial and coastal plains, subject to recurrent freshwater or low-salinity flooding.

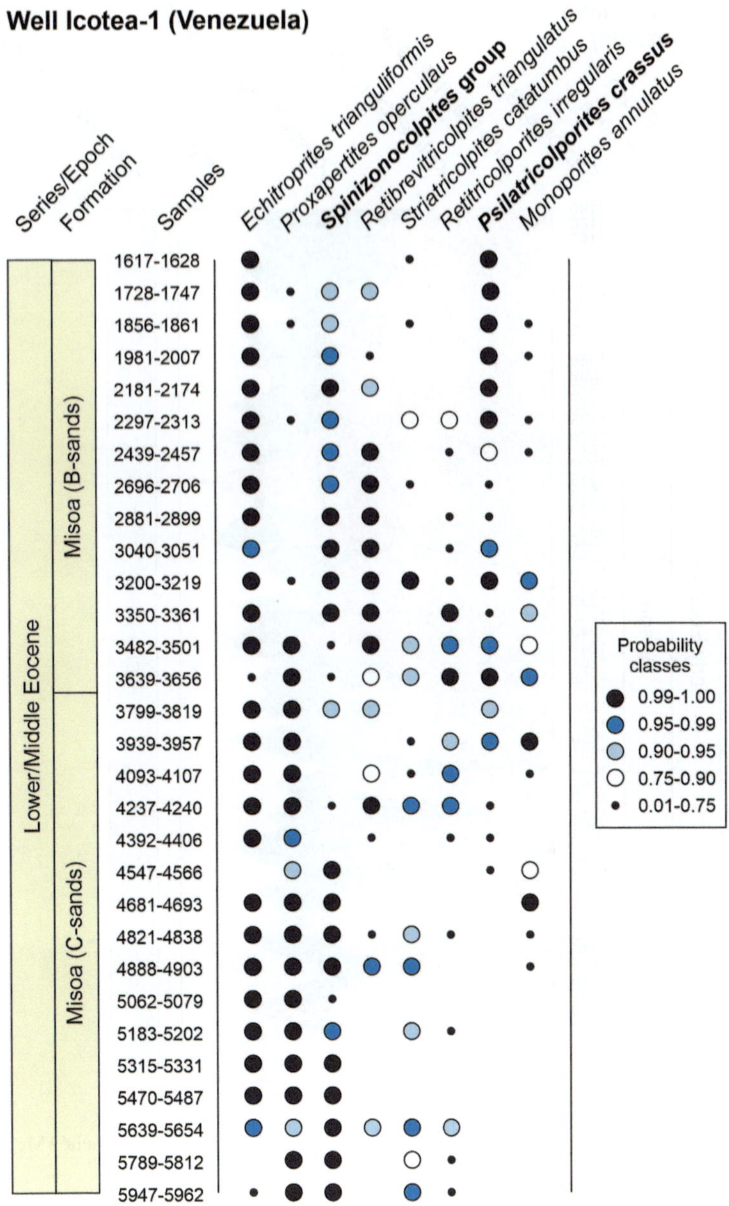

Fig. 3.9 Semiquantitative record of the most abundant Eocene palynomorphs from well Icotea-1 (Ic), in Venezuela (Fig. 3.2). Redrawn and simplified from Germeraad et al. (1968)

The Lutetian VLC interval, as depicted in Fig. 3.11, was identified by four primary ecological groupings according to Rull (1998a, b, c, d). These groupings encompassed inland forests marked by *Retitricolpites simplex* and *Retimonocolpites*

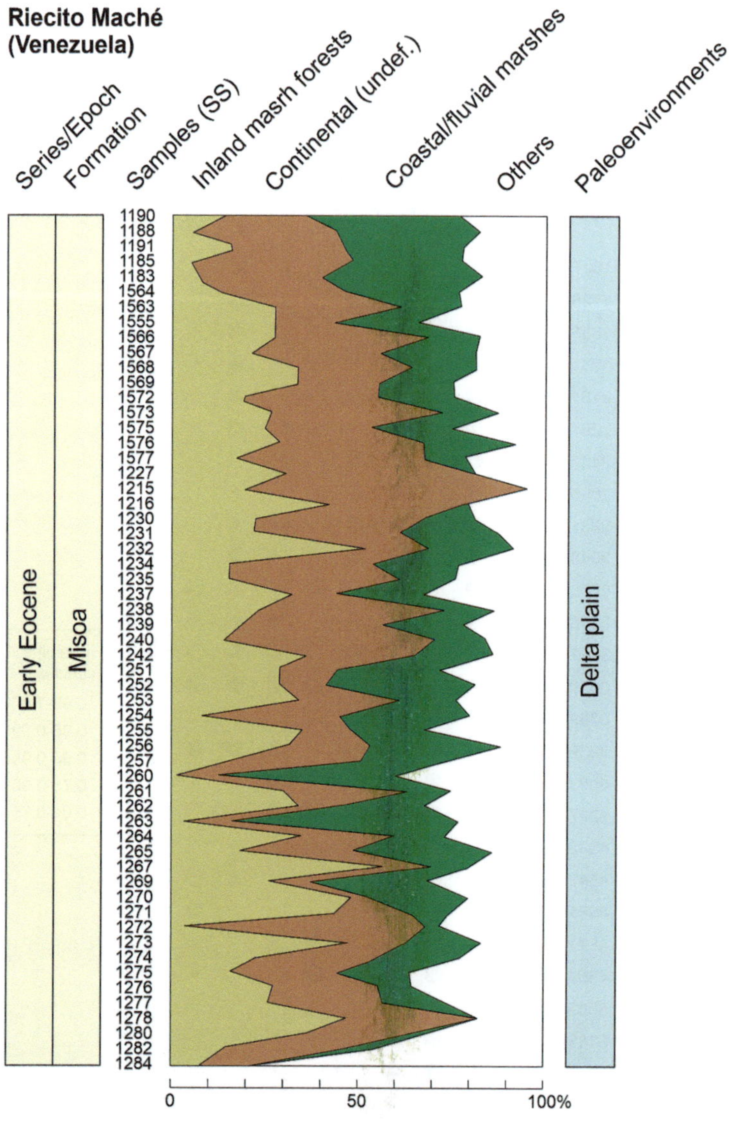

Fig. 3.10 Early Eocene palynological assemblages defined in the Riecito Maché (Mc) section, Venezuela. Redrawn and modified from Rull (1999)

fossulatus, palm-dominated back-mangrove swamps featuring *Longapertites proxapertitoides* and *Mauritiidites franciscoi*, fern-dominated back-mangrove swamps with *Deltoidospora adriennis* (*Acrostichum*) and *Verrucatosporites speciosus*, and a mangrove-forest assemblage showcasing *Lanagiopollis crassa* (*Pelliciera*) and *Spinizonocolpites echinatus*/*S. prominatus* (*Nypa*). The palynological sequence demonstrated a characteristic mangrove ecosystem with the typical elements arranged in a sea–inland zonal pattern. Specifically, it included a *Pelliciera*-dominated mangrove forest along the coastline, with *Nypa* growing either beneath or immediately behind *Pelliciera* in saline waters. Moving further inland, a back-mangrove complex emerged with a decreasing salinity gradient, featuring ferns (*Acrostichum*) in brackish-water swamps and palms (*Mauritia*) in freshwater environments. The dating of this Lutetian section relied on a palynostratigraphic zonation specific to western Venezuela, incorporating the upper part of the *Echitriporites trianguliformis* zone, the *Retitricolporites magnus* zone, and the lower part of the *Bombacacidites soleaformis* zone. According to Rull (1998b), this dating places the age of the sequence between <49 and >49 Ma.

The distinctive association observed between *Lanagiopollis crassa* (*Pelliciera*) and *Spinizonocolpites echinatus*/*S. prominatus* (*Nypa*), coupled with a significant surge in the former during the Middle Eocene, agrees with findings in other regions, such as those investigated in Icotea-1 and Prevención (Germeraad et al. 1968).

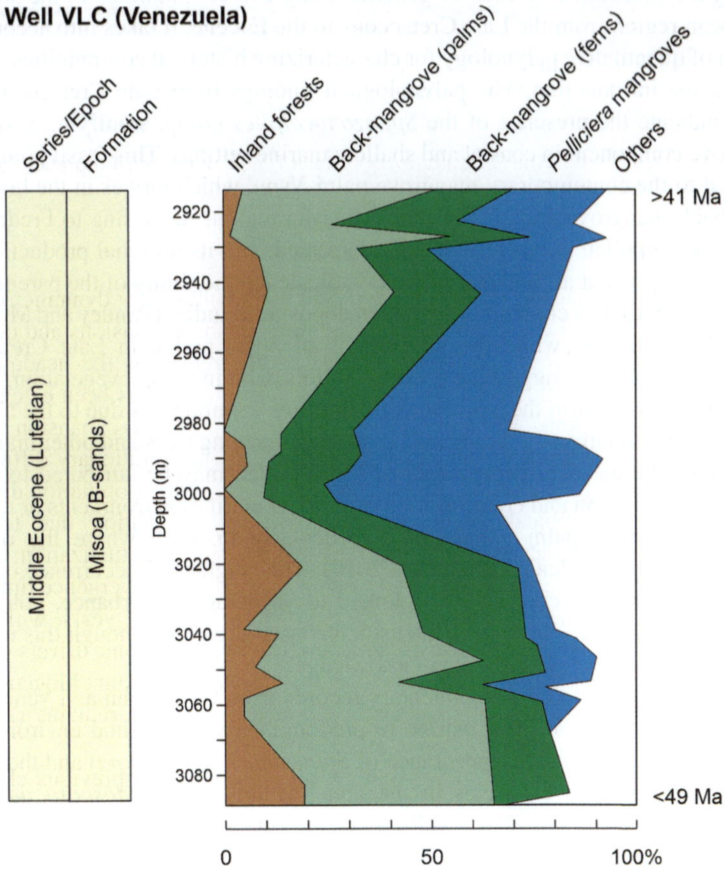

Fig. 3.11 Palynological assemblages defined in a Middle Eocene (Lutetian) interval of the VLC (VC) well from NE Lake Maracaibo (Block III). Redrawn and modified from Rull (1998b)

Likewise, in the Sagu (Sg) section, there are notable local abundance peaks for *Lanagiopollis crassa* (*Pelliciera*) and the *Spinizonocolpites group* (*Nypa*), particularly in the Middle Eocene part of the sequence (De la Parra et al. 2021). The Piñalerita section also reveals an increase in the prevalence of both morphospecies during the Early/Middle Eocene transition, as reported by Jaramillo and Dilcher (2001). Furthermore, Graham (1977, 1985) highlighted the frequent occurrence of *Pelliciera* in the Middle Eocene Chapelton (Cp) Formation in Jamaica and the Eocene Gatuncillo Formation near Alcalde Díaz (AD) in Panama (Fig. 3.2). In aggregate, this body of evidence strongly indicates a widespread presence of true mangrove communities in the Middle Eocene at a regional level.

3.2 The Oldest Caribbean Mangroves

This section delves into the examination of the previously discussed evidence concerning the potential existence of genuine mangrove communities in the southern Caribbean region from the Late Cretaceous to the Eocene. It takes into account the criteria of quantitative palynology for characterizing historical communities, as outlined in the introduction. The palynological findings from Late Cretaceous sediments indicate the presence of the *Spinizonocolpites* group, signifying a singular mangrove component in coastal and shallow marine settings. This fossil pollen type is linked to the contemporary mangrove palm *Nypa*, which thrives in the brackish-water back-mangrove zone in the Indo-Malayan region. According to Frederiksen (1985), *Nypa* pollen is typically locally dispersed, and its minimal production and buoyancy imply that a small quantity can indicate the proximity of the parent plant, a notion supported by contemporary pollen deposition studies (Pandey and Minckley 2019). Nevertheless, while the proportions of *Nypa* pollen in Late Cretaceous coastal environments suggest the presence of this palm in coastal vegetation, it does not definitively confirm the existence of mangrove communities due to the absence of fossil representatives of recognized mangrove-forming trees and other mangrove associates. The elevated percentages of *Nypa* pollen may be attributed to downstream transportation and concentration in shallow marine environments or the proliferation of *Nypa* palm populations, particularly in areas where the original vegetation has been cleared (Numbere 2019). The heightened occurrence of *Nypa* pollen in the fossil record could be linked to vegetation disturbance, potentially instigated by climatic changes or intensified coastal erosion, although this remains speculative given the current state of knowledge.

The examined Paleocene sedimentary records from Colombian and Venezuelan sites reveal a transition from marine to predominantly continental environments. This shift resulted in the disappearance of *Spinizonocolpites* (*Nypa*) and the prevalence of non-mangrove species in the area. Although *Deltoidospora adriennis* (*Acrostichum*) is recognized as a well-represented mangrove component, it is essential to recognize that, like *Nypa*, *Acrostichum* is not a mangrove-forming tree. Consequently, its presence does not accurately signify the existence of mangrove ecosystems; rather, it indicates the occurrence of coastal areas inundated by low salinity waters or freshwaters, as supported by its association with the freshwater palm *Mauritia* (*Mauritiidites*). Consequently, there is a lack of evidence for Paleocene mangrove ecosystems in the southern Caribbean region, similar to the situation in the Late Cretaceous. Of note, *Deltoidospora* is not found in Colombian locations, implying a prevalence of more inland (fluvial) environments. It is also significant that *Spinizonocolpites* and *Deltoidospora* did not coexist in the examined sections, potentially due to their distinct environmental preferences and likely influenced by frequent lateral facies changes (Jaramillo and Dilcher 2001). Furthermore, problems in identifying *Deltoidospora adriennis*, as highlighted by Frederiksen (1985), may also contribute to these observations.

During the Early Eocene, the situation resembled that of the Paleocene, with the exception of *Brevitricolpites variabilis*, identified by González Guzmán (1967) as a potential mangrove-forming tree. However, challenges arose due to the original definition of this morphospecies by Jaramillo and Dilcher (2001), the absence of botanical links to known mangrove species and the lack of other recognized mangrove components, all of which undermined this assertion. Additionally, *B. variabilis* was not consistently present in the region at the levels initially reported, and there is no additional evidence supporting its presumed role as a mangrove-forming species. Therefore, owing to uncertainties in taxonomy, a lack of botanical connections, and the limited spatiotemporal occurrence of *B. variabilis*, this fossil cannot be considered a significant element in the evolution of Caribbean mangroves. The situation underwent a shift in the Middle Eocene when quantitative pollen/spore records indicated the existence of true mangrove ecosystems. *Pelliciera* emerged as the dominant mangrove-forming tree, accompanied by *Nypa*, *Acrostichum*, and *Mauritia* as key elements reflecting a typical mangrove zonal pattern associated with a decreasing salinity gradient from marine to freshwater conditions. This represents the earliest robust evidence of mangrove ecosystems in the Caribbean. While back-mangrove elements were present in earlier sediments (Late Cretaceous to Early Eocene) along coastal areas, the absence of a predominant mangrove-forming tree species and the lack of a zonal distribution of back-mangrove elements along the coastal-inland salinity gradient made it difficult to infer the presence of typical mangrove communities using quantitative criteria before the Middle Eocene.

3.2.1 Synthesis

The available quantitative palynological evidence indicates that the initial establishment of Caribbean mangroves likely occurred in the southern part of the region, specifically in present-day Colombia and Venezuela, during the Middle Eocene (Lutetian), approximately 50–40 Ma. *Pelliciera* (*Lanagiopollis crassa*) was the first mangrove-forming tree in these coastal communities, while the palm *Nypa* (*Spinizonocolpites* spp.) grew in the understory or immediately beyond *Pelliciera*, within the tidal zone. Back-mangrove areas featured the fern *Acrostichum* (*Deltoidospora adriennis*) and the palm *Mauritia* (*Mauritiidites franciscoi*), along with other palm (*Psilamonocolpites* spp.), suggesting a decreasing salinity gradient from brackish water to freshwater. The fossil representative of *Pelliciera*, *L. crassa*, first appeared in the Early Eocene and had already expanded to other Caribbean regions by the Middle Eocene. The dispersion of Caribbean mangrove communities across the region is believed to have been facilitated by the migration of the Caribbean plate. Morphological and genomic studies have revealed two variants, A and B, within modern *Pelliciera rhizophorae*. Variant A thrives in wet zones, while Variant B thrives in areas with low rainfall. It is proposed that these variants may have diversified early in response to environmental conditions or pollination vectors (Castillo-Cárdenas et al. 2016).

The overall picture represents a change of paradigm in our comprehension of the origins of Caribbean mangroves. Contrary to earlier beliefs, it seems that these mangroves did not directly descend from purported Late Cretaceous Tethyan mangroves. Instead, they independently emerged much later, during the Middle Eocene, following the appearance of the autochthonous *Pelliciera*, the first documented mangrove-forming tree in the AEP region, which is absent in the IWP region. *Pelliciera* played a crucial role in establishing a new ecosystem by bringing together species adapted to tidal conditions (such as *Nypa* and *Acrostichum*), aiding the adaptation of others to saline and brackish-water environments, and contributing to the evolutionary emergence of post-Eocene mangrove-forming tree taxa, as discussed in subsequent chapters. The emergence of *Pelliciera* appears to have been a pivotal event, acting as a "condensation nucleus" for organizing and expanding the novel Neotropical mangrove ecosystem exclussive from the AEP region. As mentioned earlier, the presence of mangrove-forming trees is vital for the existence of mangrove ecosystems. The review underscores that *Pelliciera*, originating in the Early Eocene, dominated coastal environments in the Middle Eocene, marking the commencement of unique and characteristic mangrove ecosystems in the Caribbean. Furthermore, a distinctive zonal mangrove pattern, encompassing both the mangrove forest and a back-mangrove community gradient toward inland environments, has been identified—an observation not previously noted in the region. This new understanding challenges existing theories about the origin and evolution of mangroves, particularly those rooted in the concept of progressive regional differentiation from a presumed uniform Late Cretaceous pantropical mangrove belt. The lack of quantitative pollen records for hypothetical Late Cretaceous and Paleocene mangrove communities in the Caribbean region implies a different paleobiogeographical and evolutionary scenario than previously envisioned.

As is well-known, absence of evidence is not evidence of absence. Consequently, one should not rule out the possibility that future research might uncover substantial evidence supporting the existence of Caribbean mangroves predating the Middle Eocene. However, based on the presently available information, the aforementioned conclusions stand as the most parsimonious explanation. This novel perspective is anticipated to establish a foundation and stimulate further progress in this research domain. A compelling avenue for exploration involves extending the quantitative methodology to other regions housing mangrove ecosystems in tropical and subtropical zones, with the objective of attaining a more thorough and global comprehension of mangrove origins and evolution. In such investigations, the primary emphasis should be on identifying potential tree species responsible for mangrove formation and establishing a distinct zonal pattern of mangroves through quantitative paleoecological reconstructions. It is crucial to recognize that merely having or lacking fossil records for specific salt-tolerant species is insufficient for the identification and reconstruction of past mangrove communities.

References

Birks HJB, Birks HH (1981) Quaternary palaeoecology. E. Arnold, London

Castillo-Cárdenas MF, Sanjur O, Toro-Perea N (2016) Differences in sculpture and size of pollen grains: new morphological evidence of diversification in *Pellicicera rhizophorae*, an ancient Neotropical mangrove species. Palynology 40:302–307

De la Parra F, Pinzon D, Mantilla-Duran F et al (2021) Marine-lacustrine systems during the Eocene in northern South America – palynological evidence from Colombia. J S Am Earth Sci 108:103188

Dodd RS, Afzal-Rafii ZA (2002) Evolutionary genetics of mangroves: continental drift to recent climate change. Trees 16:80–86

Dueñas HJ (1989) Palynological associations of the Colón and Molino formations, northeast Colombia. Bol Inst Geocien Spec Publ 7:173–181

Duke NC, Lo EYL, Sun M (2002) Global distribution and genetic discontinuities of mangroves – emergning patterns in the evolution of *Rhizophora*. Trees 16:65–79

Duke NC, Meyneccke J-O, Dittman S et al (2017) A world without mangroves? Science 317:41b–42b

Ellison JC (2008) Long-term retrospection on mangrove development using sediment cores and pollen analysis: a review. Aquat Bot 89:93–104

Ellison AM, Farnsworth EJ, Merkt RE (1999) Origins of mangrove ecosystems and the marine biodiversity anomaly. Glob Ecol Biogeogr 8:95–115

Escobar M, Márquez G, Suárez-Ruis I et al (2016) Source-rock potential of the lowest coal seams of the Marcelina Formation at the Paso Diablo mine in the Venezuelan Guasare Basin: evidence for the correlation of Amana oils with these Paleocene coals. Int J Coal Geol 163:149–165

Faegri K, Iversen J, Kaland E et al (1989) Textbook of pollen analysis. Blackburn Press, Caldwell

Frederiksen NO (1985) Review of early Tertiary sporomorph paleoecology. Am Assoc Strat Palynol Contr Ser 19:1–92

Garzón S, Warny S, Bart PJ (2012) A palynological and sequence-stratigraphic study of Santonian-Maastrichtian strata from the Upper Magdalena Valley basin in central Colombia. Palynology 36(Suppl 1):112–133

Gee CT (2001) The mangrove palm *Nypa* in the geologic past of the New World. Wetl Ecol Manag 9:181–194

Germeraad JH, Hopping CA, Muller J (1968) Palynology of Tertiary sediments from tropical areas. Rev Palaeobot Palynol 6:189–348

González de Juana C, Iturralde JM, Picard X (1980) Geología de Venezuela y de sus Cuencas Petrolíferas (I and II). FONINVES: Caracas

González Guzmán AE (1967) A Palynological Study of the Upper Los Cuervos and Mirador formations (Lower and Middle Eocene), Tibú Area, Colombia. EJ Brill, Amsterdam

Graham A (1977) New records of *Pelliciera* (Theaceae/Pellicieraceae) in the Tertiary of the Caribbean. Biotropica 9:48–52

Graham A (1985) Studies in Neotropical paleobotany. IV. The Eocene communities of Panama. Ann Missouri Bot Gard 72:504–534

Graham A (1995) Diversification of Gulf/Caribbean mangrove communities through Cenozoic time. Biotropica 27:20–27

Higgs R (1996) A new facies model for the Misoa Formation (Eocene), Venezuela's main oil reservoir. J Petrol Geol 19:249–269

Hofman C-C (2002) Pollen distribution in sub-Recent sedimentary environments of the Orinoco Delta (Venezuela) – an actuo-palaeobotanical study. Rev Palaeobot Palynol 119:191–217

Jaramillo CA (2002) Response of tropical vegetation to Paleogene warming. Paleobiology 28:222–243

Jaramillo C, Dilcher DL (2001) Middle Paleogene palynology of Central Colombia, South America: a study of pollen and spores from tropical latitudes. Palaeontographica B 258:87–213

Jaramillo CA, Muñoz F, Cogoll M et al (2005) Quantitative biostratigraphy for the Cuervos Formation (Paleocene) of the Llanos foothills, Colombia: improving palynological resolution for oil exploration. In: Powell AJ, Riding JB (eds) Recent developments in applied biostratigraphy. Micropaleontol Soc Spec Publ, pp 145–159

Jaramillo C, Bayona G, Pardo-Trujillo A et al (2007) The palynology of the Cerrejón Formarion (Upper Paleocene) of northern Colombia. Palynology 31:153–189

Lo EYY, Duke NC, Sun M (2014) Phylogeographic pattern of *Rhizophora* (Rhizophoraceae) reveals the importance of both vicariance and long-distance oceanic dispersal to modern mangrove distribution. BMC Evol Biol 14:83

Mannion PD, Upchurch P, Benson RBJ et al (2014) The latitudinal biodiversity gradient through deep time. Trends Ecol Evol 29:42–50

Montaño PC, Nova G, Bayona G et al (2016) Análisis de secuencias y procedencia en sucesiones sedimentarias de grano fino: un ejemplo de la Formación Umir y base de la Formación Lisama, en el sector de Simacota (Santander, Colombia). Bol Geol 38:51–72

Moore PD, Webb JA, Collinson ME (1991) Pollen analysis. Blackwell, Oxford

Muller J (1959) Palynology of recent Orinoco delta and shelf sediments. Micropaleontology 5:1–32

Muller J (1981) Fossil pollen records of extant angiosperms. Bot Rev 47:1–140

Numbere AO (2019) Impact of invasive *Nypa* palm (*Nypa fruticans*) on mangroves in coastal areas of the Niger Delta region, Nigeria. In: Makowski C, Finkl CW (eds) Impacts of invasive species on coastal environments. Springer Nature, Cham, pp 425–454

Pandey S, Minckley TA (2019) Modern pollen-vegetation studies from the Sajnekhali Island Wildlife Sanctuary, Sundarbans, Eastern India. Palynology 43:213–222

Plaziat J-C, Cavagnetto C, Koeniguer J-C et al (2001) History and biogeography of the mangrove ecosystem, based on a critical reassessment of the paleontological record. Wetl Ecol Manag 9:161–179

Pocknall DT, Erlich RN (2020) Palynostratigraphy and lithostratigraphy of Upper Cretaceous and Paleogene outcrop sections, Mérida Andes (Maracaibo Basin), Western Venezuela. J S Am Earth Sci 104:102830

Polanía J, Urrego LE, Agudelo CM (2015) Recent advances in understanding Colombian mangroves. Acta Oecol 63:82–90

Ramon JC, Fajardo A (2006) Sedimentology, sequence stratigraphy, and reservoir architecture of the Eocene Mirador Formation, Cupiagua Field, Llanos Foothills, Colombia. Am Assoc Petrol Geol Mem 88:433–469

Romero EJ (1993) South American paleofloras. In: Goldblatt P (ed) Biological relationships between Africa and South America. Yale University Press, New Haven, pp 62–85

Rull V (1987) A note on pollen counting in Palaeoecology. Pollen Spores 29:471–480

Rull V (1998a) Biogeographical and evolutionary considerations on *Mauritia* (Arecaceae), based on palynological evidence. Rev Palaeobot Palynol 100:109–122

Rull V (1998b) Middle Eocene mangroves and vegetation changes in the Maracaibo basin, Venezuela. PALAIOS 13:287–296

Rull V (1998c) Evolución de los manglares neotropicales: la crisis del Eoceno. Interciencia 23:355–362

Rull V (1998d) Modern and Quaternary palynological studies in the Caribbean and Atlantic coasts of northern South America: a paleoecologically-oriented review. Bol Soc Venez Geól 23:5–24

Rull V (1999) Palaeofloristic and palaeovegetational changes across the Paleocene/Eocene boundary in northern South America. Rev Palaeobot Palynol 107:83–95

Rull V (2000) Ecostratigraphic study of Paleocene and Early Eocene palynological cyclicity in northern South America. PALAIOS 15:14–14

Rull V (2002) High-impact palynology in petroleum geology: applications from Venezuela (northern South America). Am Assoc Petrol Geol Bull 86:279–300

Rull V (2020) Quaternary ecology, evolution and biogeography. Elsevier/Academic Press, London

Rull V (2022) The Caribbean mangroves: an Eocene innovation with no Cretaceous precursors. Earth-Sci Rev 231:104070

Rull V, Montoya E (2014) *Mauritia flexuosa* palm swamp communities: natural or human-made? A palynological study of the Gran Sabana region (northern South America) within a Neotropical context. Quat Sci Rev 99:17–33

Rull V, Vegas-Vilarrúbia T, Espinoza N (1999) Palynological record of an Early-Mid Holocene mangrove in eastern Venezuela. Implications for sea-level rise and disturbance history. J Coast Res 15:496–504

Santos CE (2012) Palynostratigraphy of the Umir Formation, Middle Magdalena Valley Basin (MMVB), Colombia. MS diss, Louisiana State Univ

Srivastava J, Prasad V (2018) Evolution and paleobiogeography of mangroves. Mar Ecol 40:e12571

Takayama K, Tateishi Y, Kajita T (2021) Global phylogeography of a pantropical mangrove genus *Rhizophora*. Sci Rep 11:7228

Tchegliakova N (1995) Los foraminíferos de la Formación Umir (sección Quebrada La Julia): registro del Cretácico Superior cuspidal (Maestrichiano) en el Valle Medio del Magdalena, Colombia. Geol Colomb 19:109–130

Triest L (2008) Molecular ecology and biogeography of mangrove trees towards conceptual insights on gene flow and barriers: a review. Aquat Bot 89:138–154

Urrego LE, González C, Urán G et al (2010) Modern pollen rain in mangroves from San Andrés Island, Colombian Caribbean. Rev Palaeobot Palynol 162:168–182

Van der Stocken T, Carroll D, Menemenlis D et al (2019) Global-scale dispersal and connectivity in mangroves. Proc Natl Acad Sci USA 116:915–922

Villamil T (1999) Campanian-Miocene tectonostratigraphy, depocenter evolution and basin development of Colombia and western Venezuela. Palaeogeogr Palaeoclimatol Palaeoecol 153:239–275

Willard DA, Weimer LM, Riegel WL (2001) Pollen assemblages as paleoenvironmental proxies in the Florida Everglades. Rev Palaeobot Palynol 113:213–235

Woodroffe CD, Grindrod J (1991) Mangrove biogeography: the role of Quaternary environmental and sea-level change. J Biogeogr 18:479–492

Oligocene Revolution

4

Abstract

The Eocene *Pelliciera* mangroves were replaced by different mangrove communities dominated by *Rhizophora*, the precursors of modern mangroves, during the Eocene–Oligocene transition (EOT). The EO (T33.8–33.5 Ma) was characterized by relevant global tectonic and climatic disruptions that greatly influenced biotic patterns worldwide. In the Caribbean region, the EOT disruption was manifested in an abrupt cooling (3–6 °C) and sea-level fall (67 m), coinciding with a shift in mangrove dominance from the autochthonous *Pelliciera* to the allochthonous *Rhizophora*, originating in the IWP and arriving by long-distance dispersal in the Late Eocene. *Pelliciera* remained as a subordinate mangrove element and was restricted to a small equatorial patch around the Panama Isthmus, where it still thrives. It is proposed that the EOT cooling and sea-level fall could have favored the expansion of the eurythermic and vagile *Rhizophora*, which outcompeted the stenothermic *Pelliciera*, of limited dispersal ability. The survival of *Pelliciera* could have been facilitated by *Rhizophora*, which would have provided shelter against environmental stressors, such as light intensity and salinity. In this way, competence would have turned into facilitation, thus promoting coexistence by niche segregation. In this trade, *Pelliciera* could be viewed as an ecological looser, by losing its dominance, but an evolutionary winner, by surviving under generally unfavorable conditions.

Contemporary mangroves in the Caribbean exhibit marked differences in their composition compared to their counterparts from the Mid–Late Eocene, as discussed in Chap. 3. Unlike the Eocene mangroves, which featured *Pelliciera* as a primary mangrove-forming tree, this species underwent rapid extinction in most areas after the Eocene. The sole remaining representative, *Pelliciera rhizophorae*, is currently restricted to a specific region around the Panama Isthmus. In this area, it plays a secondary role within mangrove communities dominated by different tree species

V. Rull, *Origin and Evolution of Caribbean Mangroves*, Ecological Studies 252,
https://doi.org/10.1007/978-3-031-57612-6_4

(Dangremond and Feller 2014; Dangremond et al. 2015; Duke 2020). Unlike Eocene mangroves, extant Neotropical mangroves are characterized by the prevalence of mangrove trees such as *Rhizophora*, *Avicennia*, and *Laguncularia* (Tomlinson 2016). These trees come from distinct families and orders, without direct phylogenetic connections to *Pelliciera* (Table 4.1 and Fig. 4.1). To delve into the corresponding phylogenetic relationships, see, for example, Schwarzbach and Ricklefs (2000), Schwarzbach and McDade (2002), Anderberg et al. (2002), Geuten et al. (2004), Schönenberger et al. (2004), Fonseca (2021), and Maurin et al. (2021). This suggests that current Caribbean mangrove communities exhibit structural and ecological similarities to Eocene mangroves but are organized around tree species that belong to distant evolutionary lineages. Thus, the observed characteristics of habit, architecture, and physiological traits in Caribbean mangrove trees demonstrate instances of evolutionary convergence among taxonomically unrelated groups, driven by adaptation to intertidal saline environments. Consequently, the establishment of modern-like mangrove ecosystems represents a second evolutionary "surprise," marked by the emergence of new mangrove-forming trees within unexpected taxonomic groups, rather than the evolutionary development of existing primeval mangrove species.

As depicted in Fig. 4.2, the transition from Eocene mangroves to present-day Caribbean mangroves took place during the Eocene–Oligocene transition (EOT), a

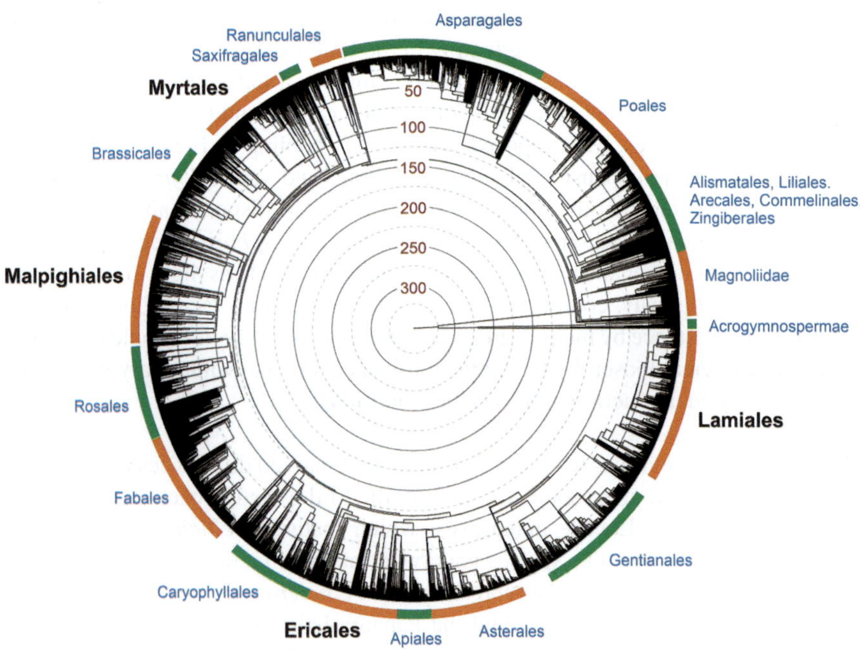

Fig. 4.1 Angiosperm phylogeny using data from GenBank and the Open Tree of Life project (Smith and Brown 2018). Black numbers are divergence ages in million years. The orders containing Caribbean mangrove-forming trees are highlighted in black bold letters (see Table 4.1 for details)

Table 4.1 Caribbean mangrove-forming trees indicating the families and orders to which they belong to highlight their phylogenetic disparity (see also Fig. 4.1)

Family (Order)	Genus	Range	Species	Caribbean	Range
Malpighiaceae (Malpighiales)	*Rhizophora*	Pantropical	6	*R. mangle*[a] *R. racemosa*	AEP (Af-Am) AEP (Af-Am)
Acanthaceae (Lamiales)	*Avicennia*	Pantropical	8	*A. germinans*[a] *A. bicolor* *A. schaueriana*	AEP (Af-Am) AEP (Am) AEP (Am)
Combretaceae (Myrtales)	*Laguncularia* *Conocarpus*	AEP AEP	1 1	*L. racemosa*[a] *C. erectus*	AEP (Af-Am) AEP (Af-Am)
Tetrameristaceae (Ericales)	*Pelliciera*	AEP	1	*P. rhizophorae*	AEP (Am)

AEP, Atlantic–East Pacific region (Fig. 2.4); Af, Africa; Am, America. Raw data from Tomlinson (2016) and Duke et al. (2017)
[a]Most abundant/characteristic species

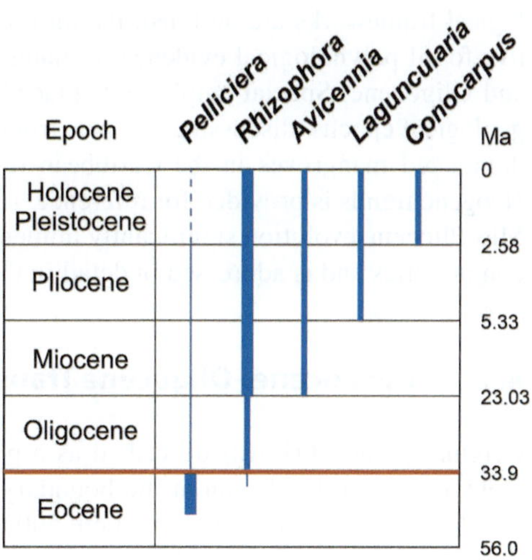

Fig. 4.2 General stratigraphic chart of fossil pollen representing the modern Caribbean mangrove trees. Redrawn from Rull (2022a). Boundary dates after Cohen et al. (2013)

geological boundary marked by significant disruptions in global environment and biogeography (Prothero and Begggren 1992; Coxall and Pearson 2007). Consequently, the development of Caribbean mangroves, and, by extension, Neotropical mangroves, has been characterized by two major unexpected and abrupt incidents: (i) the emergence of primeval Eocene *Pelliciera* mangrove communities and (ii) their subsequent decline and replacement by modern-like post-Eocene mangrove forests. After the EOT, the evolution of post-Eocene mangrove forests exhibited a trend of increasing species richness without observable extinctions, influencing the current taxonomic and biogeographical patterns of Caribbean mangroves

(Graham 1995; Rull 2023). While the first event (i) has been thoroughly addressed in Chap. 3, this chapter focuses on the second event (ii).

In this chapter, mangrove communities are reconstructed using the same community approach as in the preceding chapter. It also explores the evolution of key components of these communities, specifically the mangrove-forming tree species, through the application of molecular phylogeographical methods or spatial projections of dated phylogenies. These methods enable robust reconstructions of spatiotemporal patterns in mangrove communities, which are then compared with trends in paleoenvironmental and paleogeographical data obtained from independent proxies. The aim is to identify potential chronological correlations indicating plausible causal relationships. To enhance comprehension, the chapter is divided into three main parts. The initial part offers a global overview of major events during the EOT, establishing a broad paleoenvironmental, paleogeographical, and biogeographical framework for this specific time period. Topics covered include global climate change and its potential causes, biotic responses, and the evolution of global mangrove ecosystems, focusing on *Pelliciera* and *Rhizophora*. The second part concentrates on the Caribbean region, summarizing its physical aspects (tectonics, paleogeography, paleoclimates, paleocirculation, sea-level shifts) and biogeographical features during the EOT. Once these general physical and biological frameworks are outlined, the third part analyzes qualitative and quantitative fossil palynological evidence of mangrove evolution throughout the Eocene and Oligocene. Special emphasis is placed on the transition between these two geological epochs, discussing the shift from *Pelliciera*-dominated to *Rhizophora*-dominated mangroves in the Caribbean region. While a summary of the main Neogene trends is provided for reference, it is not extensively analyzed here. The Mio-Pliocene evolution significantly influenced present-day Caribbean mangrove communities and is addressed in detail in the subsequent chapter.

4.1 The Eocene–Oligocene Transition

Originally, the EOT was described as a period featuring rapid changes in climate and biota that began at the boundary between the Eocene and Oligocene epochs (33.9 Ma) and lasted for about 500,000 years (Coxall and Pearson 2007). Subsequent research offered a slightly adjusted perspective, situating the EOT between 34.4 and 33.7 Ma, with a duration of approximately 800,000 years (Hutchinson et al. 2021). The initial comprehensive synthesis of available evidence was presented in a foundational multidisciplinary book by Prothero and Begggren (1992). More recent updates to this overview by Coxall and Pearson (2007) and Hutchinson et al. (2021) integrated advancements in methodological approaches to paleoclimatic proxies and acknowledged the significant contributions of the Ocean Drilling Program (ODP; https://www.iodp.org/) in characterizing the EOT.

4.1.1 Climatic and Sea-Level Changes

Presently, there exists extensive and consistent evidence supporting the occurrence of global cooling during the EOT. This evidence is drawn from the analysis of oxygen isotope ($\delta^{18}O$) records of benthic foraminifera, which were obtained from deep-sea marine cores collected globally by the ODP. Numerous studies, including those by Diester-Haass and Zahn (1996), Zachos et al. (1996), Coxall et al. (2005), Coxall and Pearson (2007), Pearson et al. (2008), Coxall and Wilson (2011), Bohaty et al. (2012), Wade et al. (2012), Borrelli et al. (2014), and Langton et al. (2016), have contributed to this understanding (Fig. 4.3). The global cooling observed at the EOT signifies the shift from Cenozoic greenhouse climates, predominant on Earth until the Eocene, to the current Cenozoic icehouse climates. A noteworthy consequence of this cooling was the glaciation of Antarctica, which had remained unglaciated until the Eocene (Katz et al. 2008; Young et al. 2011). It is important to highlight that there is no comparable evidence supporting glaciation in the Northern Hemisphere, and it is widely accepted that Antarctica was the sole region to experience glaciation during the Oligocene (Zachos et al. 2001; Westerhold et al. 2020). Evidence for glaciation in the Northern Hemisphere only emerged in the Late Miocene (11–6 Ma) in Greenland, as indicated by studies conducted by Larsen et al. (1994), Helland and Holmes (1997), Maslin et al. (1998), Bierman et al. (2016), and Pérez et al. (2018). A significant intensification of glaciation is observed across Greenland, Eurasia, and North America around the Pliocene/Pleistocene boundary (2.6 Ma), as documented by Shackleton et al. (1984), Raymo (1994), Ehlers and Gibbard (2007), and De Schepper et al. (2014).

The cooling linked to the EOT played a pivotal role in establishing distinct latitudinal temperature gradients, as precursors to the current global climatic zones on Earth (Pagani et al. 2011; Straume et al. 2022). In the Eocene, there existed a low-temperature gradient, resulting in high-latitude temperatures that were 20–40 °C higher than today, with tropical temperatures only 5–10 °C above current levels (Huber and Sloan 2000; Huber 2008; Bijl et al. 2009; Huber and Caballero 2011). During that time, frost-intolerant flora and fauna were present at high latitudes (Greenwood and Wing 1995). Various continental biotic records suggest a trend toward aridification during the EOT, although it remains uncertain if this was a global phenomenon (Coxall and Pearson 2007; Hutchinson et al. 2021). In addition to climate changes, there were abrupt shifts in sea levels across the E/O boundary. The available evidence indicates a general sea-level decrease of approximately 70 meters, which is 50 meters below present levels (Miller et al. 2008; Houben et al. 2012; Wilson et al. 2013). This sea-level variation is likely linked to Antarctic glaciation, resulting in marine disconnections that influenced circulation patterns (Coxall et al. 2018; Hutchinson et al. 2019).

The cooling of the Earth's surface and the formation of ice in Antarctica are ascribed to the interaction of three main factors: a significant decrease in the overall atmospheric concentration of carbon dioxide (CO_2), the establishment of the Antarctic Circumpolar Current (ACC), and an increase in albedo caused by the expansion of the

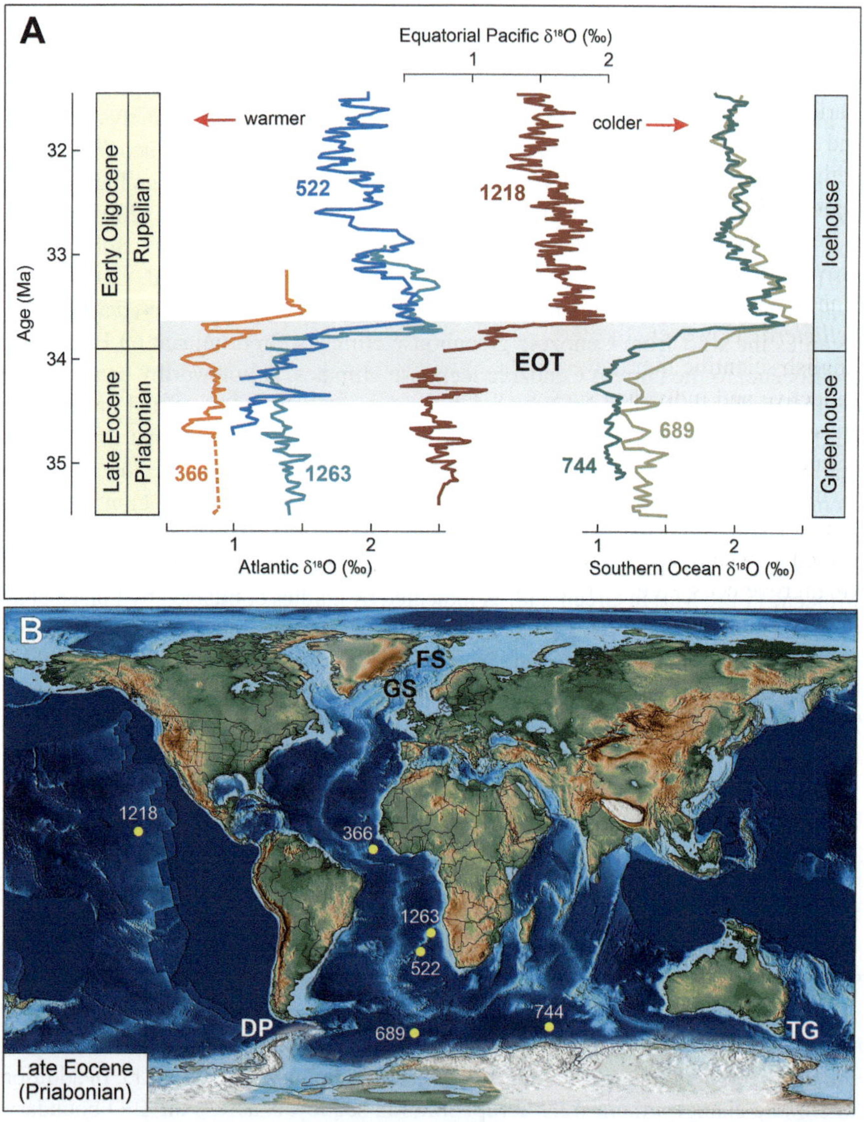

Fig. 4.3 Worldwide cooling across the Eocene–Oligocene transition (EOT). (**a**) Oxygen isotope curves from several ODP (Ocean Drilling Program) cores obtained in different oceans and different latitudes (see **b** for location). Modified from Hutchinson et al. (2021). (**b**) Late Eocene paleogeography with the location of the above ODP cores and the gateways mentioned in the text: DP, Drake Passage; FS, Fram Strait; GS, Greenland-Scotland Rise; TG, Tasman Gateway. Base map from the PALEOMAP Paleo Atlas for GPlates (https://www. earthbyte.org/paleodem-resource-scotese-and-wright-2018/)

ice sheet (DeConto and Pollard 2003; Hutchinson et al. 2021; Straume et al. 2022). During the EOT, there was a notable decline in pCO_2 levels, decreasing from around 1000 parts per million (ppm) in the Priabonian to 700–800 ppm in the Rupelian, resulting in a substantial decrease in the greenhouse effect. Various modeling approaches suggest that this CO_2 reduction could have been the primary driver of the cooling (Hutchinson et al. 2021). Additionally, the tectonic opening of Southern Ocean pathways in the Drake Passage and the Tasman Gateway (Fig. 4.3) during the Late Eocene played a pivotal role. This would have enabled the establishment of ACC circulation, isolating Antarctica thermally and expediting its glaciation (Barker and Burrell 1977; Barker and Thomas 2004). While the Drake Passage was open before the Late Eocene, it was initially too shallow to sustain the ACC (Livermore et al. 2007; Markwick 2007). The deepening of the Tasman Gateway opening, surpassing 500 m, occurred around 33.5 Ma, near the E/O boundary, making both pathways capable of supporting the ACC by 30 Ma (Hutchinson et al. 2021). Recent insights suggest that connections between the Arctic and North Atlantic through the Fram Strait and the Greenland–Scotland Rise (Fig. 4.3) may have also played a role in EOT cooling and Antarctic glaciation by regulating the southward flow of cold Arctic freshwaters (Straume et al. 2022). According to modeling results, changes in radiative forcing due to the expansion of the Antarctic ice sheet, driven by albedo effects, played a secondary role in this process (Hutchinson et al. 2021).

4.1.2 Biotic Responses

The E/O boundary marked a significant phase of biological transformations. This passage was characterized by increased rates of extinction during the Eocene and heightened radiation rates during the Oligocene. Notably, this boundary witnessed major changes in biogeography, attributed to the rapid cooling and decline in sea levels during the EOT. These environmental shifts led to the establishment of permanent icehouse conditions. Although the biotic changes during this period did not reach the magnitude of the well-known "Big Five" mass extinctions (Sepkoski (1986); Jablonski and Chaloner (1994), they had a noteworthy impact on both terrestrial and marine biotas (Coxall and Pearson 2007). The effects of these changes are evident in marine microfossils, which serve as chronostratigraphic markers. For instance, the Hantkeninidae family of planktonic foraminifera, whose last occurrence serves as a marker in the Global Stratotype Section at the E/O boundary (Premoli Silva and Jenkins 1993). Also, tropical planktonic foraminifera adapted to warmer conditions experienced extinction (Boersma and Premoli Silva 1986; Keller et al. (1992). Similar instances of biotic turnover are observable across various marine organisms, including benthic foraminifera, calcareous nannoplankton, radiolarians, dinoflagellates, diatoms, ostracodes, and shallow-marine invertebrates (Baldauf 1992; Berggren et al. 1995; Schellenberg 1998; Dockeri and Lozouet 2003; Nebelsick et al. 2005; Van Mourik and Brinkhuis 2005; Funakawa et al. 2006).

Compared to marine records, there is significant spatial variation in the turnover of continental biota during the EOT, as highlighted by Sheldon et al. (2016). This diversity is especially evident in the records of mammals and plants. Concerning mammals, a notable faunal turnover known as the "Grand Coupure" took place in Eurasia concurrently with the onset of Oligocene glaciation. This event led to the extinction of a substantial portion (up to 60%) of indigenous European species, which were then replaced by immigrants from Asia (Hooker et al. 2004). The factors contributing to these changes are attributed to a combination of climate-induced extinction, competition, and the establishment of long-distance dispersal routes resulting from sea-level regression. In Asia, the parallel turnover in mammalian populations was primarily associated with climate change (Meng and McKenna 1998). However, there is no evidence of similar occurrences in the mammalian faunas of North and South America (Prothero 2012; Woodburne et al. 2014). Evidence from these continental regions suggests diverse responses to environmental changes.

The primary characteristic of how vegetation responds to disruptions at the EOT is its varied nature across continents, as evidenced in a recent global study that utilized worldwide pollen/spore databases (Pound and Salzmann 2017). The authors attribute the diverse patterns of vegetation response to the EOT to a combination of factors including tectonics, sea-level changes, declining greenhouse gas concentrations, and more localized influences such as orogeny and precipitation patterns. During the E/O boundary, global vegetation underwent significant shifts, but the specific nature of these changes differed across continents and latitudes (Coxall and Pearson 2007). For example, in North America, there was a widespread replacement of subtropical evergreen Eocene vegetation with cold-adapted deciduous Oligocene forms, accompanied by notable regional extinctions, particularly at higher latitudes (Liu et al. 2007). The continental interior experienced increased aridity, potentially leading to the diversification of desert species (Yancey et al. 2003; Moore and Janzen 2005). In southern South America, a transition from subtropical to temperate vegetation occurred, but this shift was later attributed to long-term Eocene cooling rather than an abrupt EOT shift (Quattrocchio et al. 2013). The Neotropics saw significant floral turnover at the E/O boundary, linked to EOT cooling, with substantial evidence concentrated in and around the Caribbean region (Germeraad et al. 1968; Frederiksen 1985; Muller et al. 1987).

In Europe, the EOT brought about significant alterations in plant life, marked by a shift toward more seasonal temperate vegetation and the disappearance of tropical and subtropical species (Collinson 1992; Eldrett et al. 2009). However, the impact of the EOT varied across different regions, leading to changes such as transitions to warm-temperate forests, deciduous mixed forests and arid and seasonal biomes, or, in some cases, no noticeable shifts in vegetation. This variability depended on the specific area and is documented by Pound and Salzmann (2017) and Hutchinson et al. (2021). In Asia, the European Grande Coupure coincided with synchronous changes in both plant and animal life, involving a transition from Eocene warm-temperate forests and large-bodied perissodactyl faunas to dry-temperate forest-steppe vegetation with small-bodied rodents/lagomorphs (Sun et al. 2014). Nevertheless, regional variations, influenced by local factors like the Tibetan uplift and associated climatic changes, also played a significant role, as indicated by Jin et al. (2017) and Su et al. (2018). Australia underwent a transition from the Eocene to the Oligocene characterized by reduced diversity, heightened seasonality, and the spread of cool-temperate flora (Sluiter et al. 2022). In Antarctica, the expansion of ice sheets coincided with the replacement of evergreen forests by shrubby and tundra vegetation (Francis and Poole 2002; Galeotti et al. 2022).

In summary, EOT was characterized not only by significant environmental shifts leading to a climate resembling the modern conditions but also by a profound change in ecosystems that facilitated the emergence of present-like biotas. Both marine and terrestrial environments saw increased extinction during the EOT, followed by a recovery in the Oligocene. However, responses in continental settings varied, with biogeographical reorganizations playing a crucial role (Coxall and Pearson 2007; Hutchinson et al. 2021). Global factors such as cooling, sea-level reduction, and decreased atmospheric CO_2, along with local influences like tectonic shifts and changes in topography, especially affecting precipitation patterns, are attributed to the biotic changes during the EOT. While not universally applicable, an increase in aridity or precipitation seasonality has been suggested in diverse continental records, particularly in interior regions, based on paleontological evidence.

4.1.3 Mangroves and the EOT

This section analyzes the present distribution of mangrove tree genera found in the Caribbean and investigates the phylogeographical patterns that may have influenced these biogeographical trends, with a specific focus on the EOT. While the genera *Rhizophora* and *Avicennia* have pantropical distributions, their species display disjunct distributions in the AEP and IWP regions. The remaining three genera—*Pelliciera*, *Laguncularia*, and *Conocarpus*—are monotypic and restricted to the AEP region, as outlined in Table 4.1 and Fig. 2.5. This review places emphasis on *Pelliciera* and *Rhizophora* as the primary mangrove-forming trees in the Caribbean during the Eocene and Oligocene, respectively (Graham 1995; Rull 1998b). The subsequent appearance of other genera (*Avicennia*, *Laguncularia*, and *Conocarpus*) during the Mio-Pleistocene is discussed in Chap. 5.

4.1.3.1 Fossil and Biogeographical Evidence

Prior to recent advancements in molecular phylogeographical research, hypotheses about biogeography and evolution primarily depended on theoretical inference, including concepts like center of diversity and center of origin, as well as fossil evidence. As advanced in the former chapter, the AEP–IWP polarity was initially explained through either dispersalist or vicarianist perspectives. Dispersalists suggested that mangroves originated in the high-diversity IWP region and then dispersed to the AEP region (Van Steenis 1962). In contrast, vicarianists proposed that mangroves evolved during the Late Cretaceous along the coasts of the continuous Tethys Sea and diversified later, coinciding with the closure of this sea by the African barrier (McCoy and Heck 1976). A more comprehensive review, incorporating macrofossil and microfossil evidence, lent support to the vicariance hypothesis (Ellison et al. 1999). Subsequently, various authors presented differing perspectives on the timing of mangrove origin and the subsequent divergence into the AEP and IWP regions, spanning from the Cretaceous to the Oligocene (e.g., Sowunmi 1986; Ellison et al. 1999; Plaziat et al. 2001; Duke 2017; Srivastava and Prasad 2015, 2018). The EOT environmental disruption, proposed as a significant factor influencing the initial split between IWP and AEP biogeographical mangrove regions (Plaziat et al. 2001; Srivastava and Prasad 2018), as well as the replacement of *Pelliciera* by *Rhizophora* mangroves in the Neotropics (Rull 1998b), was suggested based on fossil evidence. However, specific explanations for these events remain elusive, and as a result, no definitive causal connections have been established between EOT global disruption and the evolution of mangroves using fossil and biogeographical evidence.

4.1.3.2 Molecular Phylogenetics and Phylogeography

Recent progress in molecular phylogenetics and phylogeography provides valuable insights into the development of current biodiversity and biogeographical patterns in mangroves. The investigation of molecular phylogenetics in mangrove tree species, including Caribbean representatives like *Rhizophora* and *Avicennia*, began in the late 1980s and has significantly expanded since the 2000s (review by Triest 2008). While a detailed examination of these findings is beyond the scope of this paper, we will briefly outline key discoveries related to the mangrove-forming tree genera in the Caribbean before and after the E/O boundary. Notably, *Pelliciera* and *Rhizophora* played pivotal roles in this transition and will be subjected to more in-depth analysis.

Pelliciera, a genus comprising a single species (but see Duke 2020), is characterized by a limited geographical distribution and lacks close phylogenetic relatives in mangrove ecosystems worldwide. This presents a challenge in establishing phylogeographical relationships with evolutionary significance for these ecological communities. Previous phylogenetic studies on *Pelliciera* have primarily focused on its systematic aspects, aiming to determine its taxonomic placement within the family and the order to which it belongs (Anderberg et al. 2002; Bremer et al. 2002; Geuten et al. 2004; Schönenberger et al. 2004). Furthermore, molecular phylogenetic data have been utilized to investigate potential trends in incipient speciation within the extant *P. rhizophorae* (Castillo-Cárdenas et al. 2014, 2015).

The *Rhizophora* situation stands out with notable distinctions, showcasing higher diversity, a global presence, and clear biogeographical differences between the IWP and AEP mangrove regions (refer to Fig. 2.5). The existing biogeographical patterns in this genus, combined with the ancient age of its oldest fossils dating back to the Eocene (Graham 2006), position the phylogeographical trends of *Rhizophora* as a valuable tool for comprehending the evolution and historical biogeography of mangroves on a global scale. According to Xu et al. (2017), mangrove members in the Rhizophoraceae family, including IWP genera like *Bruguiera*, *Ceriops* and *Kandelia*, along with the widely distributed *Rhizophora*, diverged from their non-mangrove counterparts in the same family around 55 Ma during the Early Eocene, coinciding with the Paleocene/Eocene Thermal

Maximum. The study also indicates that the *Rhizophora* genus emerged approximately 39 Ma during the Late Eocene, and the AEP species *R. mangle* diverged from its IWP counterparts, *R. apiculata*, *R. mucronata*, and *R. stylosa*, at an unspecified point in the Neogene. It is essential to highlight that the phylogeny is calibrated with Late Eocene fossil records (Graham 2006), which introduces a challenge in directly comparing with the fossil record due to potential circularity issues.

Takayama et al. (2021) propose that the genus *Rhizophora* most likely originated in the IWP region during the Eocene and subsequently expanded globally by the Late Eocene. The dispersion is thought to have taken place through the Tethys seaway, potentially following a southern route around Africa to reach the AEP region. This migration is estimated to have occurred between 50 Ma (Early Eocene) and 23 Ma (Oligo/Miocene). The research also suggests that the split between *Rhizophora* species in the AEP and IWP regions happened around 11 Ma (Late Miocene). This divergence was influenced by the closure of the Tethys and a cooling period that restricted movement across the Indian and Atlantic Oceans through the southern African corridor. Subsequent diversification within the AEP and IWP regions is proposed to have taken place later, facilitated by global cooling during the Mio-Pliocene. These findings agree with earlier molecular phylogenetic studies supporting the IWP origin of *Rhizophora* and its subsequent spread to the AEP. However, previous studies placed the initial IWP–AEP diversification event in the Eocene, occurring between 50 and 34 Ma, and the intra-IWP diversification at the Oligocene/Miocene boundary (Lo et al. 2014). According to these earlier studies, explaining the current biogeography of *Rhizophora* requires a combination of vicariance and dispersal, as neither explanation alone is adequate. Building on these results and considering the available fossil record, Duke et al. (2017) proposed two main radiations within *Rhizophora*. The first radiation, happening in the Oligocene (35–25 Ma), is linked to the closure of the Tethys, marking the end of pantropical interchange. The second radiation, responsible for Late Miocene (~15 Ma) intra-IWP and intra-AEP diversifications, coincided with the widening of the North Atlantic and the collision between Australia and Southeast Asia.

The potential impact of EOT cooling and the associated decline in sea levels on the evolution of mangroves is rarely addressed in these phylogenetic and phylogeographical investigations. Only Lo et al. (2014) recognized that the ancestral *Rhizophora*, once thriving along the Tethys seaway, Mediterranean and Arabian coasts, and Europe, might have faced extinction in the late Tertiary due to significant cooling and environmental changes. This was highlighted in EOT-related studies by Pagani et al. (2005) and Liu et al. (2009). As a result, the potential influence of global disruptions during EOT on mangrove evolution remains largely unexplored in phylogeographical literature. Conventional explanations in the historical biogeography of mangroves tend to concentrate on shifts in dispersal pathways and barriers influenced by continental drift (Dodd and Afzal-Rafii 2002; Dodd et al. 2002; Duke et al. 2002; Triest 2008; Duke et al. 2017; Van der Stocken et al. 2019).

4.2 The Caribbean Region During the EOT

Before delving into the core subject, it is crucial to bear in mind that this book addresses the Caribbean region comprehensively. This encompasses all the coastlines from the northern Greater Antilles to the southernmost part of South America, as well as from the eastern Lesser Antilles to western Central America (refer to Fig. 1.1). The inclusion of presently inland locations, especially from South America, is indispensable because these areas were once part of proto-Caribbean coastlines. Over time, they have distanced themselves from the sea due to the highly dynamic nature of coastal changes in the region. It is noteworthy that all locations with palynological information available in the literature have been taken into consideration, with no predetermined selection criteria. The geographical variation among different regions, such as the higher number of South American locations compared to the fewer insular sites, merely reflects differences in the intensity of research efforts and the availability of data.

4.2.1 Tectonic Setting

From a geological standpoint, the Caribbean region is notable for the presence of the Caribbean microplate, situated between the North American and South American plates, as well as the Pacific plates of Cocos and Nazca (Fig. 4.4). According to current theories, the Caribbean plate originated in the eastern Pacific and subsequently migrated eastward to its present location following the Triassic–Jurassic breakup of Pangea (210–140 Ma) and the ensuing seafloor spreading between the North American and South American plates. This spreading gave rise to the Cretaceous (140–90 Ma) proto-Caribbean seaway (Pindell and Dewey 1982; Pindell 1990; Pindell et al. 2006; Mann et al. 2006; Pindell and Kennan 2009). In the Late Cretaceous (80–75 Ma), the advancing front of the Caribbean plate moved northeastward, forming the Greater-Antilles arc that collided with the Bahamas carbonate platform. This collision, occurring during the Paleocene (60–44 Ma), led to a change in plate motion from northeastward to eastward (Pindell and Dewey 1982; Pindell et al. 1988; Mann et al. 1995). This directional shift marked the initiation of the Cayman Trough and as the plate continued its eastward movement during the Miocene (44–14 Ma). The former Aves Arc disappeared giving place to a remnant known as the Aves Ridge. Concurrently, the Lesser Antilles Arc was formed (Aitken et al. 2011; Neill et al. 2013; Allen et al. 2019). For the most recent and comprehensive information supporting this model, consult the reviews by Romito and Mann (2020) and Mann (2021).

4.2.2 Paleogeography and Paleocurrents

The alterations in paleogeography are linked to tectonic developments and play a crucial role in reconstructing historical migration and dispersal routes, essential

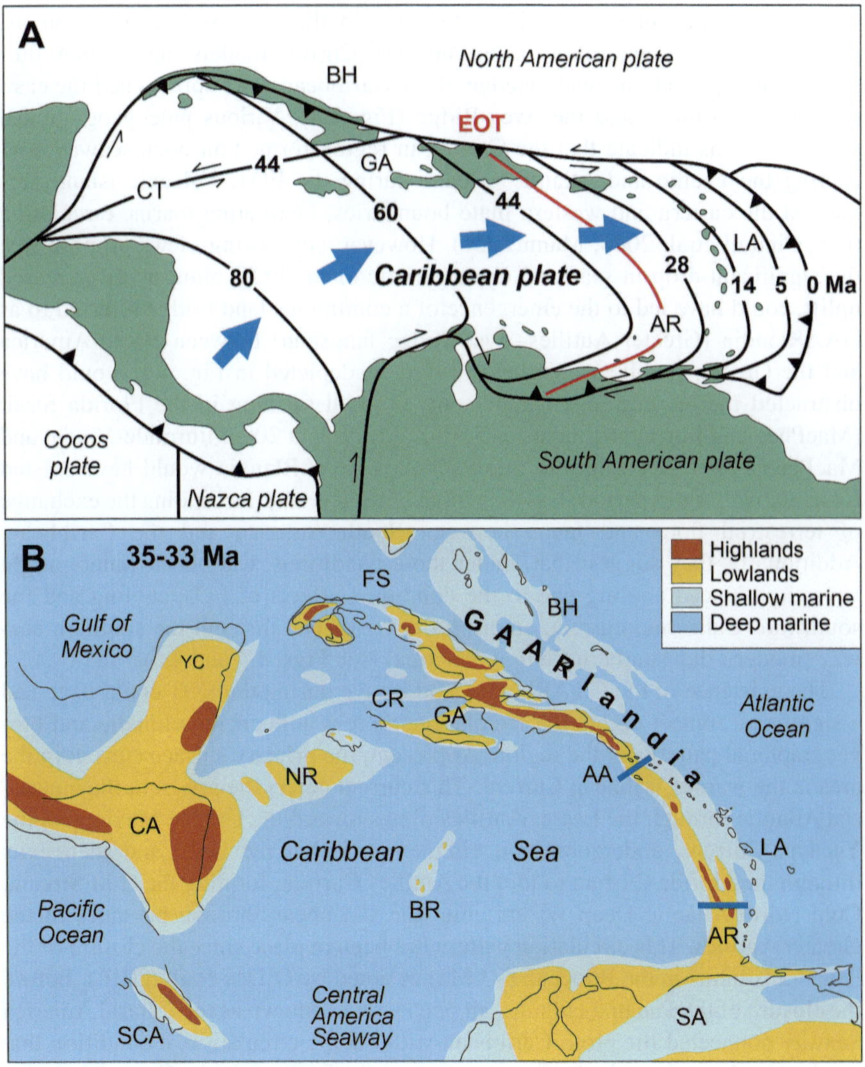

Fig. 4.4 (**a**) Late Cretaceous to present dynamics of the Caribbean Plate. Black lines indicate the position of the leading edge in the Late Cretaceous (80 Ma), Middle Paleocene (60 Ma), Middle Eocene (44 Ma), Middle Oligocene (28 Ma), Middle Miocene (14 Ma), Early Pliocene (5 Ma), and the present (0 Ma). The red line indicates the location of the advancing front during the EOT. Modified from Xie et al. (2010). (**b**) Late Eocene/Early Oligocene Caribbean paleography according to the GAARlandia hypothesis. AA, Anguilla–Antigua passage; AR, Aves Ridge; BH, Bahamas carbonate platform; BR, Beata Ridge; CA, Central America; CR, Cayman Ridge; CT, Cayman Trough; FS, Florida Strait; GA, Greater Antilles; LA, Lesser Antilles; NR, Nicaraguan Ridge; SA, South America; SCA, southern Central America; YC, Yucatán. Redrawn from Iturralde-Vinent (2006) and Iturralde-Vinent and MacPhee (1999)

for understanding changes in biogeography. In this context, we focus on the paleogeographical conditions during the EOT. Current models suggest that, during this time period, the leading edge of the Caribbean plate approached the eastern Greater Antilles and the Aves Ridge (Fig. 4.4). Various paleogeographical reconstructions indicate that the Caribbean region formed an open seaway connecting the Pacific and Atlantic Oceans during the EOT. Volcanic island arcs marked the eastern and western plate boundaries, facilitating marine circulation (e.g., Pindell et al. 2006; Mann 2021). However, contrasting views propose that the significant drop in sea-level characteristic of the EOT, along with increased uplift, could have led to the emergence of a continuous land bridge referred to as GAARlandia (Greater Antilles–Aves Ridge landspan) between South America and the Greater Antilles. This land bridge, as depicted in Fig. 4.4, would have obstructed marine currents, leaving only a small opening in the Florida Strait (MacPhee and Iturralde-Vinent 1995; Iturralde-Vinent 2006; Iturralde-Vinent and MacPhee 1999). According to these scholars, GAARlandia would have existed for a relatively short period (less than three million years), facilitating the exchange of terrestrial flora and fauna between South America and the Caribbean. Additionally, they suggest that, under these conditions, Caribbean waters might have experienced cooling due to the combined effects of EOT cooling and the southward California current, entering the Caribbean through the southern seaway (modern-day southern Central America) (see Figs. 4.4 and 4.5).

The existence of the GAARlandia land bridge during the EOT could have had a significant impact on marine circulation, thereby influencing climatic and biogeographical patterns in the region. At present, the primary surface current in the area is the warm Caribbean Current. This current enters the sea from the equatorial Atlantic through the Lesser Antilles, flows toward the Gulf of Mexico via the Yucatan Current, undergoes loop circulation within the Gulf, and then exits through the Florida Current to join the Antilles Current, forming the Gulf Stream. Cold North Atlantic Deep Waters enter the Caribbean through passages in the Greater Antilles. This circulation pattern has been in place since the closure of the Panama Isthmus in the Pliocene (~3 Ma) as noted by O'Dea et al. (2016). Before the closure of the Panama Isthmus, an open seaway known as the Central America Seaway connected the proto-Caribbean with Pacific circulation, a condition that persisted at least since the Cretaceous. During the EOT, when the Central America Seaway was open, the presence or absence of GAARlandia could have led to two main circulation modes. Without GAARlandia, the proto-Caribbean would have been connected to both the Atlantic and the Pacific, allowing for circulation between them. On the other hand, if GAARlandia did exist, it would have obstructed Atlantic circulation, leaving the Caribbean connected solely to Pacific currents. Another scenario involving the presence of GAARlandia includes the potential existence of marine passages between the Greater and Lesser Antilles, known as the AA passage (located between the present Anguilla and Antigua banks), and the Aves Ridge (AR) passage. These passages could have facilitated some connection between Caribbean and Atlantic circulation (Cornée et al. 2021; Garrocq et al. 2021).

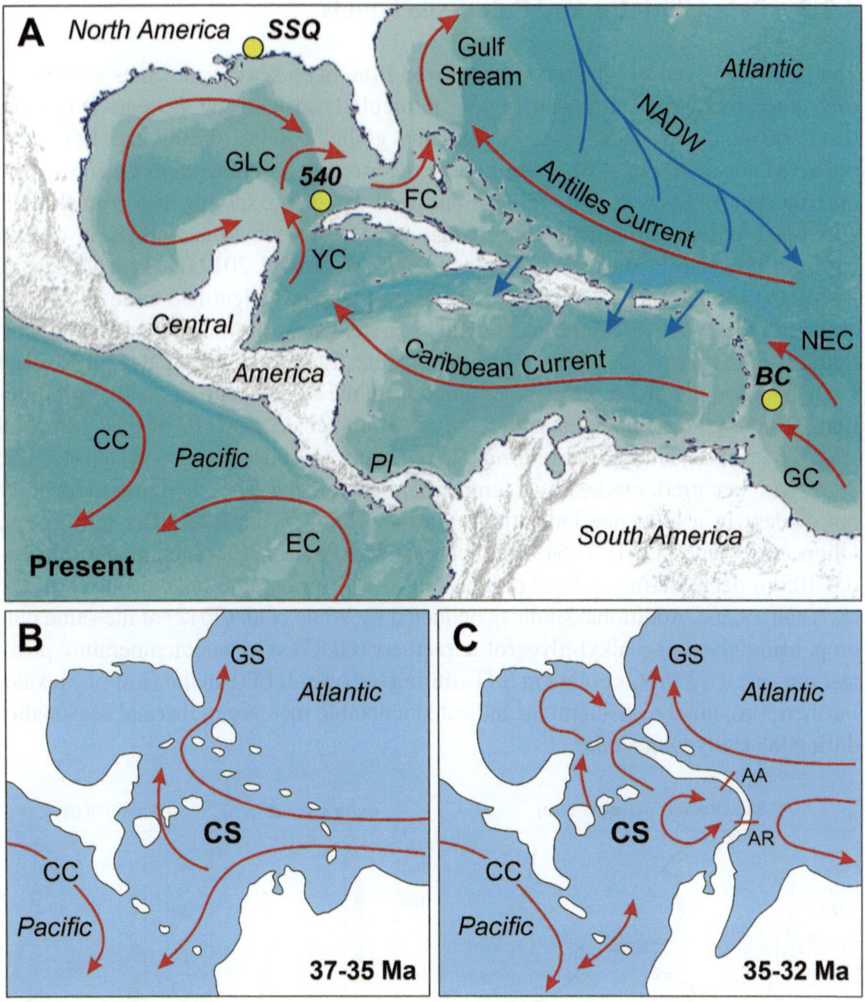

Fig. 4.5 (**a**) Present-day marine circulation in the Caribbean-Gulf of Mexico region. Red arrows represent superficial currents, and blue arrows correspond to deep currents. CC, California current; EC, equatorial countercurrent; FC, Florida Current; GC, Guiana Current; GLC, Gulf Loop Current; NADW, North Atlantic Deep Waters; NEC, North Equatorial Current; YC, Yucatan Current. Yellow dots mark the location of cores and outcrops with EOT paleoclimatic records (Fig. 4.6): BC, Bath Cliff outcrop; SSQ, St. Stephens Quarry outcrop; 540, DSDP Site 540 (marine core). (**b–c**) Late Eocene to Early Oligocene paleocurrents with (**c**) and without (**b**) GAARlandia. AA, Anguilla–Antigua passage (Cornée et al. 2021); AR, Aves Ridge passage (Garrocq et al. 2021); CS, Caribbean Sea. (**a**) Modified from Amador-Castro et al. (2021) and Öğretmen et al. (2020). (**b** and **c**) Redrawn from Iturralde-Vinent (2006)

4.2.3 Paleoclimates and Paleo-sea Levels

The limited oxygen isotope records retrieved from deep-sea marine cores and terrestrial outcrops in the Caribbean region, particularly around the Eocene–Oligocene (E/O) boundary, provide evidence that the global cooling during the EOT also impacted this area (Fig. 4.6). Among these records, the Saint Stephens Quarry (SSQ) outcrop stands out as a highly reliable and well-dated site, serving as a crucial reference point for reconstructing global changes in EOT paleoenvironments (Keigwin and Croliss 1986; Katz et al. 2008; Miller et al. 2008; Wade et al. 2012; De Lira Mota et al. 2020). In the analysis of the SSQ record, researchers have identified three key events around the EOT (33.8–33.5 Ma), indicating a stepwise transition rather than a continuous one (Katz et al. 2008). The initial event, EOT-1, occurred at 33.8 Ma and involved a cooling of 2.5 °C, as determined from the Mg/Ca ratio of benthic foraminifera. This cooling coincided with a minor 30-meter decrease in relative sea level (Fig. 4.6). Subsequently, after returning to pre-event conditions, a second step (EOT-2; 33.63 Ma) occurred, marked by a temperature change that is challenging to quantify but evident in a sequence indicating a sustained fall in sea level. The third event, Oligocene event 1 (Oi-1; 33.545 Ma), characterized by a 2 °C cooling and a substantial 105-m decrease in sea level (equivalent to a 67 m eustatic fall), occurred in the early Oligocene. Additional studies conducted by Wade et al. (2012) at the same outcrop, using glycerol-dialkyl-glycerol-tetraethers (GDGTs) as paleotemperature proxies, suggested a 3–6 °C cooling in SSTs during the overall EOT in the Gulf of Mexico. Furthermore, these measurements indicated a notable increase in thermal seasonality during the Early Oligocene.

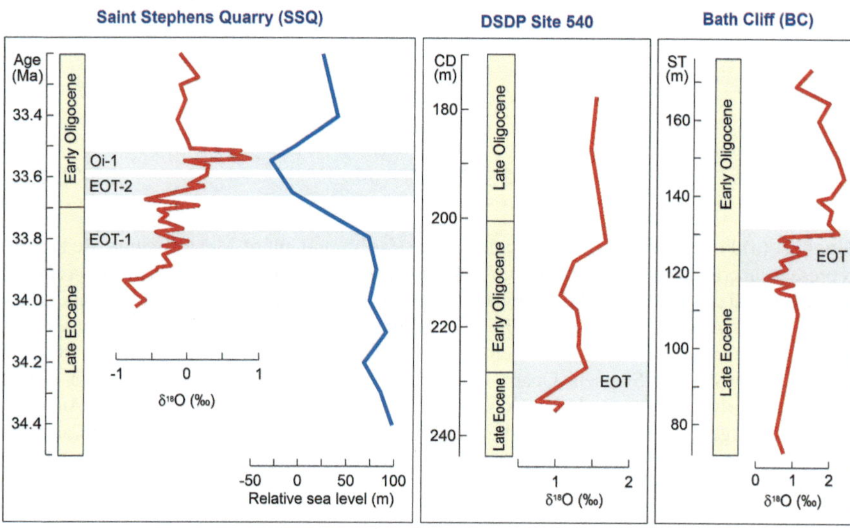

Fig. 4.6 EOT oxygen isotope (red) and sea-level (blue) records from the Caribbean region (see Fig. 4.5 for location). CD, core depth; ST, section thickness. Modified from Belanger and Matthews (1984), Saunders et al. (1984), and Katz et al. (2008)

4.2.4 Biogeography

Roncal et al. (2020) recently investigated the origin of the Caribbean biota, with a specific focus on plants. They proposed four key hypotheses to shed light on the historical evolution of terrestrial Caribbean biogeographical patterns. These hypotheses encompass (i) Late Cretaceous vicariance, (ii) GAARlandia colonization, (iii) transoceanic dispersal, and (iv) in situ speciation. The Late Cretaceous vicariance (LCV) hypothesis posits that the Caribbean biota originated on a proto-Antillean land mass positioned at the forefront of the migrating Caribbean plate, present-day Central America, approximately 80 to 70 Ma. Allopatric speciation would have occurred due to eastward migration and gradual fragmentation, resulting in the current Antilles. Nevertheless, critics challenge the existence of proto-Antillean land masses before the Middle Eocene (40 Ma). An alternative hypothesis, presented by Iturralde-Vinent and MacPhee (1999), is the GAARlandia (GAA) hypothesis, which suggests that Antilles colonization occurred in the Late Eocene/Early Oligocene, originating from South America through the GAARlandia land bridge. Long-distance dispersal (LDD) from nearby continents, particularly South America, has been a favored explanation for many decades. Water currents from South America to the Greater Antilles during the Cenozoic period are considered the primary dispersal mechanisms, as proposed by Darlington (1983), Hedges et al. (1992), and Regalado et al. (2018). Lastly, the in situ speciation (ISS) hypothesis has been introduced to explain the origin and maintenance of the region's high species richness and endemism. According to this hypothesis, species evolved locally in the Caribbean, contributing to the unique biodiversity observed in the area (Losos and Schluter 2000).

All the aforementioned theories find some empirical backing in the geographical distribution and phylogeographical patterns of specific taxonomic groups. However, there is no conclusive evidence favoring one hypothesis over the others. Recent studies employing molecular phylogenetic methods suggest certain general trends. A meta-analysis encompassing the entire Caribbean region estimated the colonization time of almost 90 endemic plant lineages, concluding that plants have consistently colonized the Antilles over the last 60 million years (approximately the Cenozoic) from continental America, particularly Central and South America (Roncal et al. 2020). This contradicts the LCV hypothesis. Notably, more than 80% of these colonization events occurred after the EOT, challenging the GAA hypothesis as a robust explanation for the origin of the Caribbean biota. Over half of the species examined colonized the Caribbean in the last ten million years, aligning with a previous study by Antonioli et al. (2015), which noted an upswing in Neotropical dispersal events during the same period, thereby supporting the LDD hypothesis. Despite these findings, the meta-analysis did not establish a correlation between colonization time and the number of species in each endemic clade, undermining the ISS hypothesis. Based on these outcomes, Roncal et al. (2020) endorse the LDD hypothesis and emphasize the potential role of Central America as a source for Caribbean flora, with almost 40% of the studied lineages originating from this continental area.

As the focus of this chapter is the EOT, the presence or absence of GAARlandia emerges as a pertinent concern. GAARlandia holds significance not only as a potential pathway for the dispersion of terrestrial biota, a commonly emphasized aspect in current biogeographical models, but also as a barrier influencing marine dispersal between the Atlantic and Pacific oceans. Moreover, it would have played a pivotal role in altering the internal circulation of the Caribbean (Fig. 4.5). Despite numerous studies highlighting island colonization in the origin of Caribbean biota, this review underscores the importance of continental coasts as typical habitats for mangrove ecosystems. It is crucial to note that verifying the existence of GAARlandia necessitates straightforward geological and geophysical evidence, which is currently unavailable. While biogeography has significantly contributed to the concept of GAARlandia, it can only provide indirect evidence.

MacPhee and Iturralde-Vinent (1995) introduced the term GAARlandia to explain the diversity of Early Miocene land mammals in the Greater Antilles, specifically concentrating on Cuba and Puerto Rico. Before this, the authors had suggested the potential existence of a continuous Oligocene subaerial pathway for land vertebrate colonization between South America and the Greater Antilles, though they had not yet coined the term GAARlandia (MacPhee and Iturralde-Vinent 1994). According to their findings, during the Late Eocene–Oligocene, the ancestral clades of Antillean mammals were widely spread across a persistent positive topographic feature in the Caribbean basin, identified as GAARlandia. This feature included regions such as "Cuba, Hispaniola, Puerto Rico, Virgin Islands, Aves Island, and some islands on the northern fringe of South America." The disruption and fragmentation of GAARlandia since the Late Oligocene were suggested to be responsible for the development of modern insular mammal faunas through vicariance and local extinction processes. At that time, the GAARlandia concept played a crucial role in explaining the distribution patterns of various land faunal groups in the Greater Antilles (Borhidi 1985; MacPhee and Iturralde-Vinent 1994, 1995, 2000, 2005; Iturralde-Vinent and MacPhee 1999). Recently, this model has found support from molecular phylogeographical studies across diverse taxonomic groups, including amphibians, mammals, insects, arachnids, and plants (Fritsch 2003; Dávalos 2004; Van Ee et al. 2008; Moen and Wiens 2009; Alonso et al. 2012; Bacon et al. 2012; Deler-Hernandez et al. 2018; Pederneiras et al. 2018; Tong et al. 2019).

As previously mentioned, the direct geological and geophysical evidence supporting the existence of GAARlandia is inconclusive (Ali 2012). Recent seismic findings indicating submerged environments in the Anguilla–Antigua banks and the Aves Ridge during the EOT challenge the idea of a continuous emerged land bridge from a physical perspective (Cornée et al. 2020, 2021; Philippon et al. 2020; Garrocq et al. 2021) (Fig. 4.4). This seismic evidence has been used to counter the GAARlandia hypothesis (Ali and Hedges 2021). Despite this, the debate continues, with some researchers suggesting that, even in the absence of a continuous terrestrial pathway, biotic connection could have occurred through stepping stone dispersal (Pennington and Dick 2004; Muellner-Riehl and Rojas-Andrés 2022). In a hypothetical fragmented GAARlandia scenario, uncertainties persist regarding the mechanisms of intra-Caribbean circulation and the colonization of continental coasts, which are crucial factors for the water dispersal of mangrove trees. Given the current state of knowledge, the biogeography and evolution of mangroves during the EOT should consider two general paleogeographical scenarios, referred to here as GAAR (with the variant frag-GAAR) and non-GAAR, as illustrated in Fig. 4.5.

4.3 The EOT Caribbean Mangroves

Following an overview of the tectonic, paleogeographical, paleoclimatic, and biogeographical context of the Caribbean, this section introduces qualitative and quantitative pollen records essential for investigating the significant disruption faced by mangrove communities during the EOT. The majority of the locations under scrutiny are found in the current countries of the southern Caribbean region, specifically Colombia and Venezuela, as previously specified. In the northern sector, only two sites in Puerto Rico (Lares; Lr) and Jamaica (Chapelton; Cp) contain relevant fossil mangrove records for this study. Conversely, the other two sites in Cuba and Haiti (Hispaniola) showcase pollen assemblages representing inland environments. The oldest of these inland Greater Antillean records (Mariaguán; Mg), dating back to the Middle Eocene, indicates the presence of terrestrial vegetation with North American floral affinities flourishing under warm-temperate to tropical climates (Graham et al. 2000). The Haiti record (Mirebalais; Ms), from the Late Miocene, portrays montane pine and cloud forests, also with North American affinities, characteristic of warm and rainy climates (Graham 1990). These records were utilized to challenge the potential occurrence of arid climates (Graham 2000) and to assert that proto-Cuba/Hispaniola islands were part of the North American plate, not directly linked to South America during the Cenozoic (Graham et al. 2000).

This discussion explores the existing evidence on paleocommunities of mangroves in the Caribbean, focusing on three significant time intervals: the Eocene, the Oligocene, and the Neogene (Miocene and Pliocene) (Fig. 4.7). The Eocene mangroves are not extensively discussed here, as they have already been covered in Chap. 3. The attention now turns to the EOT and the shift from *Pelliciera* to *Rhizophora* mangroves. It is important to note that certain Oligocene records are occasionally labeled as Oligo-Miocene in the literature due to factors like insufficient chronological resolution or the transitional nature of the studied sequence. In this review, these records are grouped under the Oligocene category, and only those explicitly identified as Miocene and Pliocene in the original references are placed in the Neogene category. While the primary focus of this paper is on the EOT, acknowledging the Neogene is crucial, especially considering that *Rhizophora* mangroves experienced their peak development during this period. Although a thorough examination of Neogene mangroves in terms of the ecological and evolutionary trends of their components and the communities they form is not presented here, recognizing their significance is deemed essential. A more in-depth analysis of Neogene mangroves is available in Chap. 5, as previously mentioned.

It is important to highlight that a number of *Pelliciera* records in Europe, Africa, and North America have been documented by different researchers (Germeraad

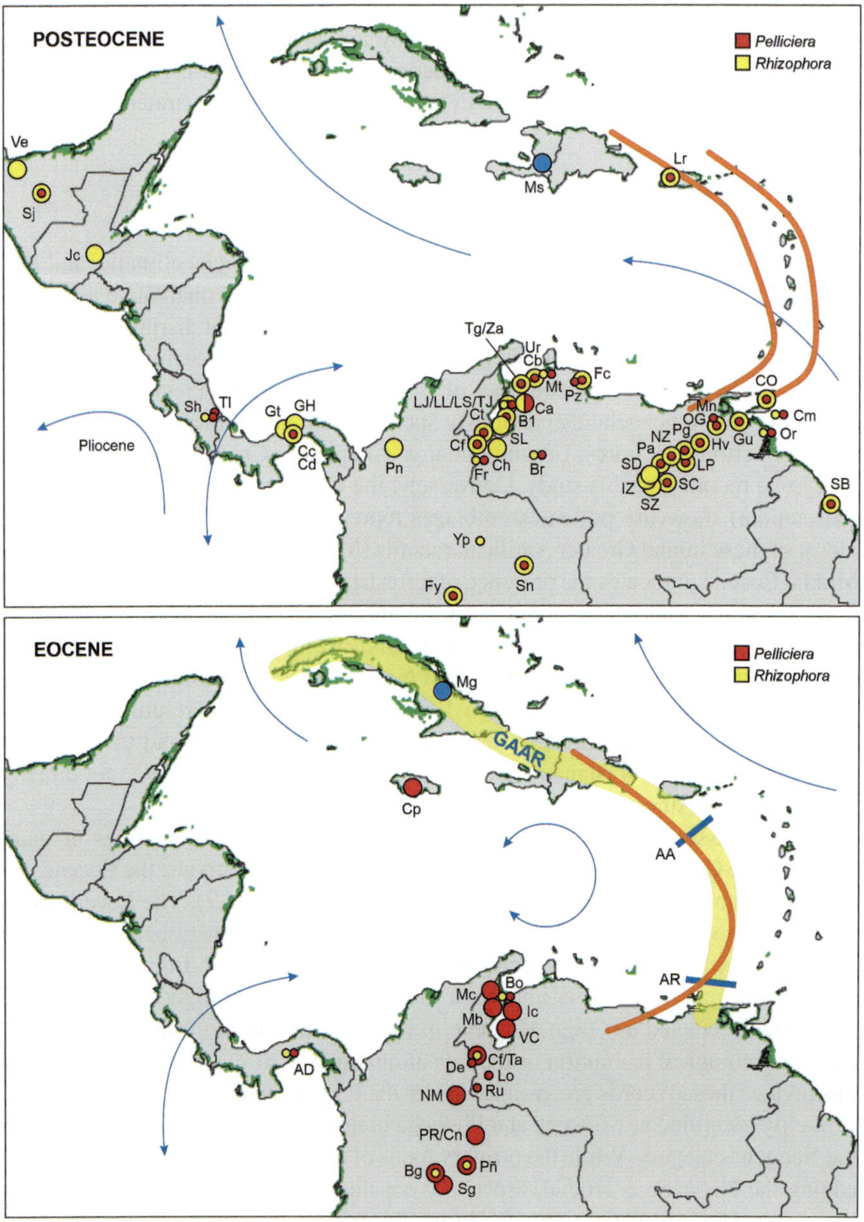

Fig. 4.7 Eocene and post-Eocene *Pelliciera* (red) and *Rhizophora* (yellow) pollen records from the Caribbean region (site abbreviations as in Table 2.4 and Fig. 2.8; pollen data from Table 4.2). Large dots indicate dominance and small dots indicate presence or low abundance. The localities with non-mangrove pollen assemblages are in blue. Orange lines mark the position of the leading edge of the Caribbean plate, and the yellow band corresponds to the approximate location of GAARlandia (AA and AR are the proposed marine connections). Paleocurrents are represented by blue lines

et al. 1968; Machin 1971; Elsik 1974; Châteauneuf 1980; Gruas-Cavagnetto et al. 1980). Frederiksen (1985) pointed out the uncertain accuracy of these identifications. Notably, *Pelliciera* is not included in the comprehensive review on the Late Cretaceous and Cenozoic history of North American vegetation (Graham 1999), which extensively covers mangrove dynamics and evolution. As a result, this review focuses on confirmed *Pelliciera* records while acknowledging the potential for new interpretations based on future discoveries.

The importance of the representativeness of pollen in relation to its plant origins is a crucial consideration in discussions regarding fossil records. Examining the link between the amount and spatial distribution of sedimentary pollen and their source plants often requires drawing parallels with contemporary patterns. For example, in the case of *Rhizophora*, modern analogs show that its pollen dominates Caribbean mangrove communities due to its high production and efficient dispersal. This pollen is particularly abundant in local mangrove environments, gradually decreasing inland but maintaining its prevalence in shallow marine sediments along coastal areas with mangroves (Rull 2022b). As a result, *Rhizophora* pollen serves as a reliable indicator of coastal and shallow marine settings. On the other hand, the situation with *Pelliciera* is not as well understood due to its limited distribution and the scarcity of pollen dispersal studies. However, being an insect-pollinated species with large, heavy pollen grains (Duke 2020), *Pelliciera* pollen tends to settle locally, near the communities of its parent plant. Therefore, even in small quantities, the presence of *Pelliciera* pollen can be regarded as a reliable indicator of the existence of in situ stands of this tree.

4.3.1 Eocene

The Eocene records reveal a distinct spatial arrangement primarily concentrated along the southern Caribbean margin, with the exception of a single record from Jamaica (Cp) (Fig. 4.7). It is noteworthy that all these records were situated west of the suggested GAARlandia emplacement during the Eocene, which would be consistent with the possible existence of this landmass. In this context, *Pelliciera* mangroves may be seen as geographically isolated from Atlantic circulation, falling under the influence of Pacific mangroves regarding dispersal and colonization. Recent research underscores coastal currents as the primary agents for mangrove propagule dispersal (Van der Stocken et al. 2019) (Fig. 4.8), and GAARlandia would have impeded the continuity of Atlantic coastal environments. Eocene plate tectonic models universally propose a volcanic island arc along the western margin of the Caribbean plate that was open to Pacific circulation (Romito and Mann 2020; Mann 2021). In the GAARlandia scenario, this could have significant implications for mangrove dispersal, confining it to Pacific colonization and impeding the potential expansion of *Pelliciera* mangroves into the Atlantic.

There are three objections to this viewpoint. First, it is contended that GAARlandia's existence was relatively short-lived, specifically confined to the last less than three million years of the Eocene epoch as indicated by Iturralde-Vinent (2006). In contrast,

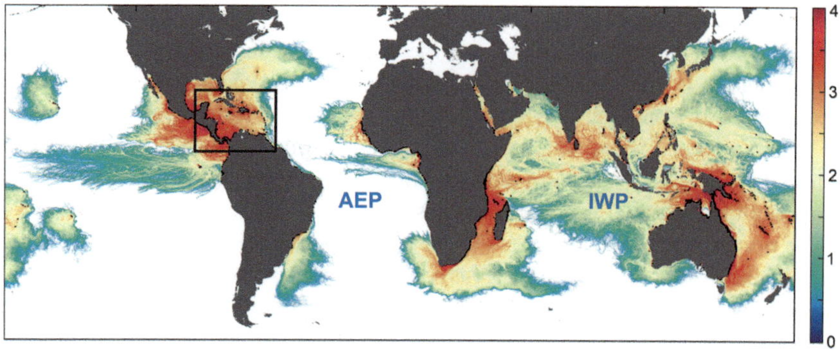

Fig. 4.8 Modeled dispersal probabilities (red, high; blue, low; white zero) for present-day mangroves modeled by combining ocean currents, and the flotability and viability of mangrove propagules. The Caribbean area is marked by a black box. Modified from Van der Stocken et al. (2019)

Pelliciera mangroves originated in the Early Eocene and flourished in the Middle Eocene (Rull 2022a), a period when the Atlantic connection was open through the present Lesser Antilles (Fig. 4.5). This suggests that these mangroves had ample opportunities for transatlantic dispersal. Second, the absence of *Pelliciera* mangroves in areas outside the Caribbean before the Middle Eocene casts doubt on the possibility of these mangroves originating from external sources. The lack of evidence implies that mangroves more likely evolved in situ rather than through immigration from the Pacific (Rull 2022a). Third, the argument that the closure of GAARlandia was incomplete, allowing some degree of Atlantic influence through the AA and AR passages (Cornée et al. 2021; Garrocq et al. 2021), suggests that the impact of GAARlandia as a dispersal barrier on the biogeographical patterns of Eocene *Pelliciera* mangroves may have been limited, except for a relatively brief period in the Late Eocene.

In the Eocene, *Rhizophora* exhibited a limited presence in the Caribbean region, being predominantly absent and sporadically observed. Graham (1977) previously noted its minor role in mangroves through qualitative records, an observation supported by our analysis of qualitative and quantitative data, summarized in Fig. 4.7. Among the 17 reported Eocene records, *Rhizophora* was absent or not mentioned in 11 cases, present in 5 instances, and relatively common (up to 10%) in only 1 location in Panama (Alcalde Díaz; AD). This variability in distribution suggests distinct biogeographical patterns at both regional and global scales. As discussed earlier, phylogeographical studies propose that *Rhizophora* likely originated in the IWP region and subsequently dispersed to the AEP during the Eocene, using the Tethys seaway for migration (Takayama et al. 2021). According to these findings, *Rhizophora* would have reached the Caribbean via the Atlantic Ocean. Current studies on mangrove dispersal pathways indicate that the Atlantic, due to its intercontinental distance, prevailing currents, and the prolonged viability of floating propagules, acts as a barrier to mangrove dispersion, including *Rhizophora*, whose propagules can remain viable for over a year in saltwater (Rabinowitz 1978). However, in the Eocene, the Atlantic was less extensive, and coastal dispersal through the former pantropical Tethys Sea was still possible (Fig. 4.3). Considering that the potential GAARlandia barrier, if present, existed only briefly close to the EOT, the theoretical possibility of *Rhizophora* dispersing from the IWP to the Caribbean through the Atlantic pathway remained viable for most of the Eocene.

One of the most notable examples of Eocene mangroves supported by quantitative data is well documented in VLC (VC), situated in Venezuelan Lake Maracaibo. In the Middle Eocene (Lutetian), *Pelliciera* emerged as the predominant mangrove-forming tree, coexisting with the palm *Nypa* (now limited to IWP; Pocknall et al. 2022). Notably, *Rhizophora* was conspicuously absent during this period (Rull 1998b) (Fig. 3.11). Comparable quantitative data is available from other locations, such as Paz del Rio (PR) and Piñalerita (Pñ) in Colombia, as well as Icotea (Ic) and Rubio Road/Río Lora (Ru/Lo) in Venezuela (Germeraad et al. 1968; Jaramillo and Dilcher 2001; Pocknall and Erlich 2020). While extensively analyzed in Chap. 3, this information is briefly reiterated here for further comparisons.

4.3.2 Post-Eocene

During the Oligocene and the Neogene, *Rhizophora* and *Pelliciera* broadened their geographical range beyond the proto-Caribbean Sea, reaching the northernmost plate boundaries in the Greater Antilles, Central America, and the Atlantic Guianan coasts (Fig. 4.7). The expansion of these mangrove trees, however, displayed distinctive characteristics. While the spread of *Rhizophora* was accompanied by overall population increases, establishing it as the predominant species in mangrove communities, the expansion of *Pelliciera* involved significant population reductions, relegating this tree to a subordinate role in mangrove vegetation. In 11 post-Eocene records (Lr, SB, OG, Cb, B1, Pa, Pg, NZ, Cd, Cd, Jc), *Rhizophora* comprised 80–100% of pollen assemblages, whereas in 12 others (Ca, Ct, LS, LJ, LP, SZ, SZ, Cc, Hv, Gu, Za, Ur), *Rhizophora* represented 40–70% of pollen counts (Table 4.2 and Fig. 4.9). The decline of *Pelliciera* was evident, with its pollen frequency

Table 4.2 Eocene, Oligocene, and Neogene records of the major mangrove-forming trees of the Caribbean region, indicating their relative abundance according to the original papers (+ present, – absent)

Site	Age	*Pelliciera*	*Rhizophora*	*Avicennia*	*Laguncularia*
SB	Pliocene–Pleistocene	–	≤90%	≤50%	–
Cm	Late Pliocene	+	+	–	–
Or	Pliocene–Pleistocene	+	+	+	–
Tl	Pliocene	+	+(?)	–	+
Mt	Pliocene	+	+	–	–
Jc	Pliocene	–	≤100%	–	–
Br	Miocene–Pliocene	+	+	–	–
Gt	Middle Pliocene	–	≤80%	–	–
Ur	Miocene–Pliocene	+	≤70%	–	–

(continued)

Table 4.2 (continued)

Site	Age	*Pelliciera*	*Rhizophora*	*Avicennia*	*Laguncularia*
Ms	Miocene-Pliocene	–	–	–	–
Ve	Late Miocene	–	≤90%	–	≤56%
Za	Late Miocene	+	≤50%	–	–
Gu	Late Miocene	+	≤70%	–	–
Hv	Late Miocene	+	≤60%	–	–
Cc	Middle Miocene	–	≤60%	–	–
Yp	Middle Miocene	–	+	–	–
TJ	Early Miocene	+	+	–	–
Sh	Early Miocene	–	≤5%	–	–
GH	Early Miocene	–	Common	–	–
SZ	Earle Miocene	–	≤60%	–	–
SC	Early Miocene	+	≤60%	–	–
IZ	Early Miocene	–	≤20%	–	–
Cd	Early Miocene	≤3%	70–90%	–	–
Sn	Early–Middle Miocene	+	Abundant	–	–
NZ	Early–Middle Miocene	+	≤80%	+	–
LP	Early–Middle Miocene	+	≤40%	+	–
Pg	Early–Middle Miocene	≤10%	≤90%	+	–
Pa	Early–Middle Miocene	+	≤90%	+	–
SD	Early–Middle Miocene	–	≤30%	–	–
LJ	Early–Middle Miocene	+	≤60%	–	–
LS	Early–Middle Miocene	+	≤40%	–	–
LL	Early–Middle Miocene	+	≤30%	–	–
Tg	Miocene	+	≤20%	–	–
Pz	Miocene	≤10%	–	–	–
CO	Miocene	+	Abundant	–	–
B1	Miocene	≤3%	≤80%	–	–
SL	Miocene	–	Abundant	–	–
Cb	Miocene	+	≤80%	–	–
Ct	Miocene	≤10%	≤40%	–	–
OG	Miocene	+	≤80%	–	–
Sj	Oligocene–Miocene	≤25%	>95%	–	–
Mn	Oligocene–Miocene	+	–	–	–
Ch	Oligocene–Miocene	–	Common	–	–
SB	Oligocene–Miocene	≤5%	≤90%	–	–
Pn	Oligocene–Miocene	–	Abundant	–	–
Fy	Oligocene–Miocene	+	Abundant	–	–
Ca	Oligocene–Miocene	≤25%	≤50%	–	–
Fr	Oligocene–Miocene	+	+	–	–
Lr	Oligocene	≤2%	≤95%	–	–
Pz	Oligocene	≤10%	–	–	–
Ct	Oligocene	+	Common	–	–

(continued)

Table 4.2 (continued)

Site	Age	*Pelliciera*	*Rhizophora*	*Avicennia*	*Laguncularia*
Tg	Oligocene	+	–	–	–
Fc	Oligocene	≤10%	Abundant	–	–
Cf	Oligocene	+	Common	–	–
Tg	Oligocene–Miocene	+	–	–	–
Cn	Late Eocene/Early Oligocene	≤47%	–	–	–
De	Eocene–Oligocene	+	–	–	–
Bg	Late Eocene/Early Oligocene	≤15%	+	–	–
Cf	Late Eocene	Com	+	–	–
AD	Middle–Late Eocene	≤11%	≤10%	–	–
Ta	Middle–Late Eocene	Common	+	–	–
PR	Middle Eocene	Abundant	–	–	–
VC	Middle Eocene	≤60%	–	–	–
Cp	Middle Eocene	Common	–	–	–
Sg	Middle Eocene	Common	–	–	–
Ic	Middle Eocene	Abundant	–	–	–
Mg	Middle Eocene	–	–	–	–
Pñ	Middle Eocene	Common	+	–	–
Bo	Middle Eocene	+	+	–	–
Ru	Early–Middle Eocene	+	–	–	–
Lo	Early–Middle Eocene	+	–	–	–
Mb	Eocene	Abundant	–	–	–
NM	Eocene	Abundant	–	–	–
Mc	Early Eocene	Common	–	–	–

Abbreviations and references as in Table 2.4 and Fig. 2.8

consistently below 10%, except in one instance (Ca), where it reached 25% of the assemblage. Qualitative and semiqualitative records indicated the presence of *Pelliciera* in 30 cases and its absence or lack of mention in 14 cases. The overall trend aligns with a widespread expansion of Caribbean mangroves linked to the replacement of *Pelliciera* by *Rhizophora* as the dominant tree. Figure 4.9 provides illustrative examples of the shift in dominance from *Pelliciera* (Eocene) to *Rhizophora* (post-Eocene). This expansion and shift in dominance of mangroves extended beyond the Caribbean region, impacting the Neotropical region, as discussed in Chap. 6.

Opportunities for dispersal were present in both the Atlantic and Pacific regions through volcanic island arcs, contributing to the spread of mangroves after the Eocene epoch. However, the lack of a similar expansion in Caribbean mangroves during the Eocene, despite comparable marine circulation conditions (except for the brief GAARlandia interval), prompts inquiry. The explanation may be related to a shift in dominance during the EOT and the specific ecological requirements of the

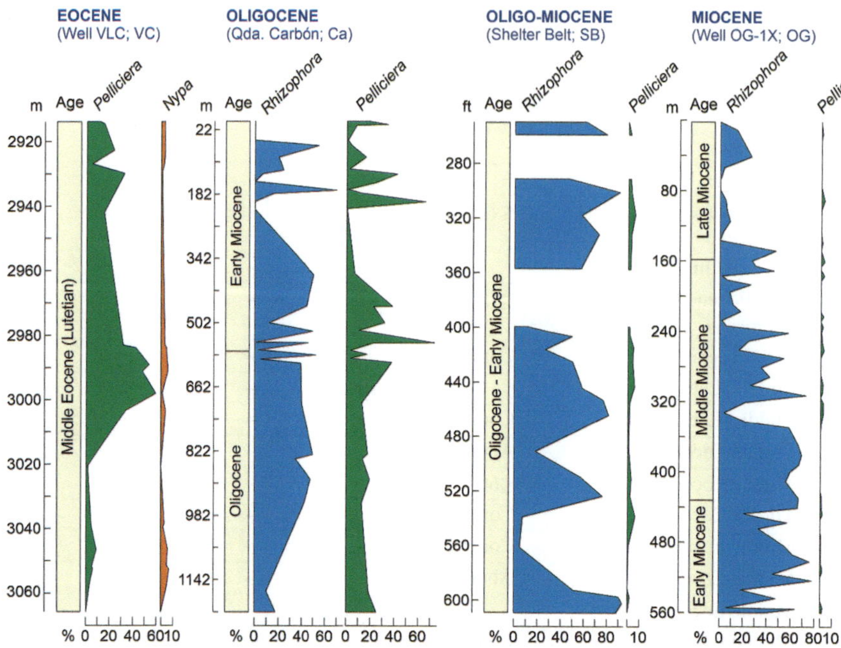

Fig. 4.9 Simplified pollen diagrams representing the gradual Eocene to Miocene *Pelliciera* to *Rhizophora* dominance shift. Depth scales in meters (m) or feet (ft), depending on the case. Assembled from Van der Hammen and Wijmstra (1964), Lorente (1986), and Rull (1998a)

species involved. Recognizing the dispersal capabilities of *Rhizophora* and *Pelliciera* propagules is crucial. *Rhizophora* propagules, able to float in saltwater for several months and remain viable for over a year, differ from *Pelliciera* propagules, which can float for just 1 week and remain viable for up to 70 days (Rabinowitz 1978). This difference may explain the expansion of Oligocene *Rhizophora* mangroves compared to Eocene *Pelliciera* mangroves. However, dispersal capacity alone does not fully clarify why *Pelliciera* also expanded in the post-Eocene alongside *Rhizophora*. One possible explanation is ecological facilitation, where one species shelters another, enabling survival in the face of environmental stress, predation or competition (Boucher et al. 1982; Callaway 1995; Stachowicz 2001; Bruno et al. 2003). Recent studies on modern mangroves in Central America lend support to this possibility, as discussed below.

The extant *Pelliciera rhizophorae* exhibits a strong sensitivity to fluctuations in both light intensity and salinity. Increased levels of these environmental stressors lead to higher mortality rates, reduced photosynthesis, and constrained growth. Conversely, when this species grows in shaded conditions, it demonstrates resilience to elevated salinity levels, indicating that light intensity plays a pivotal role as the primary limiting factor for *P. rhizophorae*. Consequently, establishing itself in areas with an open canopy becomes challenging for this species, and it typically thrives in the understory beneath the canopy of other tree species. In Central

America, *Rhizophora mangle* often provides this canopy, being a species more tolerant to environmental stressors, as illustrated by Dangremond et al. (2015). Previous theories, such as those proposed by Jiménez (1984) and Graham (1977), suggested that interspecific competition with *Rhizophora* could be a potential cause for the post-Eocene decline of *Pelliciera*. However, the findings of Dangremond et al. (2015) contradict this idea, proposing that *Rhizophora* may have actually supported the survival of *Pelliciera* by offering protection against adverse environmental factors. Alternatively, a shift from competition to facilitation may have occurred, which is not uncommon under the influence of both biotic and abiotic stress, as evidenced by contemporary ecological studies globally (He et al. 2013).

Regarding abiotic environmental factors, temperature, precipitation, and shifts in sea level, have been proposed as potential factors contributing to the decline in *Pelliciera* (Fuchs 1970; Graham 1977, 1995). Any effort to explain causation should be rooted in the ecology of these species, which is currently unknown and can only be addressed through comparisons with their extant counterparts. Such comparisons, a standard practice in paleoecological reconstructions, depend on the principle of niche conservatism over time. This principle proves particularly useful for ecosystems with long-standing existence, like mangroves (Wiens and Graham 2005; Hadly et al. 2009; Wiens et al. 2010).

Climatically, *Rhizophora* exhibits eurytopic (euryclimatic) characteristics, thriving in diverse conditions encompassing a broad spectrum of temperatures and precipitation levels in various climates, including tropical, subtropical, arid, and temperate regions (Fig. 4.10 and Table 4.3). In contrast, *Pelliciera* is stenotopic (stenoclimatic) and confined to a specific area within equatorial climates characterized by elevated temperatures, substantial precipitation, and low seasonality. Modern studies on *P. rhizophorae* typically report average temperatures of ~27 °C (with a minimum of ~18 °C and a maximum of ~33 °C) and total annual precipitation ranging from 1650 to 3050 mm (Castillo-Cárdenas et al. 2015; Dangremond and Feller 2014; Dangremond et al. 2015). In such conditions, a temperature decrease such as EOT cooling is anticipated to exert a more significant impact on a stenothermic species such as *Pelliciera* compared to a more eurythermic one such as *Rhizophora*. In theory, *Pelliciera* mangroves could have adapted (genetically) to the altered climates or migrated toward equatorial latitudes along the Pacific coasts. However, there is no evidence of such adaptations in the fossil record. Instead, severe declines in populations and local extinctions are observed at most sites (Fig. 4.7), likely attributed to the combined effects of stenothermy and limited dispersal capability. In contrast, *Rhizophora* not only adjusted to cooler Oligocene climates but also expanded its geographical range and population size, establishing dominance in mangrove communities due to its greater phenotypic plasticity, evident in its broader environmental tolerance and enhanced dispersal capacity. The emergence of new *Rhizophora* mangrove ecosystems may have provided a refuge for the survival of *Pelliciera*, as previously proposed.

The impact of EOT cooling on increased aridification or heightened seasonality in the Neotropics remains uncertain. Nevertheless, *Rhizophora* seems well adapted

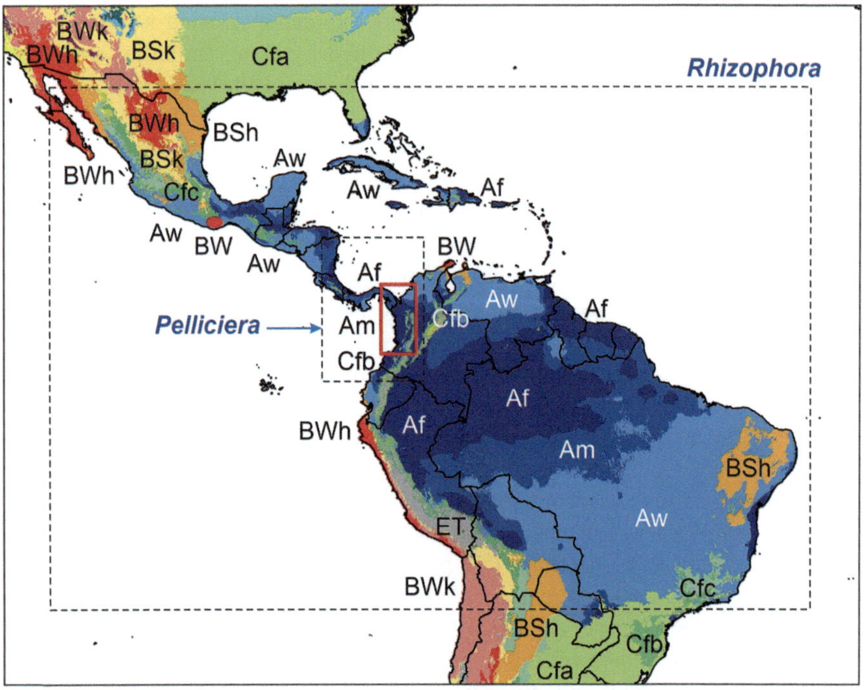

Fig. 4.10 Latitudinal and longitudinal extent of *Rhizophora* and *Pelliciera* in the Neotropics, in relation to the Köppen–Geiser climatic types (Table 4.3). The Chocó region is marked by a red box. Base map from Peel et al. (2007)

Table 4.3 Main parameters of the Köppen–Geiger climatic classification represented in Fig. 4.10 (Kottek et al. 2006; Peel et al. 2007)

Type	Description	Criteria
Af	Tropical, rainforest	Tc ≥ 18, Pd ≥ 60
Am	Tropical, monsoon	Tc ≥ 18, not Af, Pd ≥ 100-MAP/25
Aw	Tropical, savanna	Tc ≥ 18, not Af, Pd < 100-MAP/25
BWf	Arid, desert, hot	MAP<5Pt, MAT≥18
BWk	Arid, desert, cold	MAP<5Pt, MAT<18
BSh	Arid, steppe, hot	MAP≥5Pt, MAT≥18
BSk	Arid, steppe, cold	MAP≥5Pt, MAT<18
Cfa	Temperate, no dry season, hot summer	Th > 10, 0 < Tc < 18, Th ≥ 22
Cfb	Temperate, no dry season, warm summer	Th > 10, 0 < Tc < 18, not a, T10 ≥ 4
Cfc	Temperate, no dry season, cold summer	Th > 10, 0 < Tc < 18, not a/b, T10 < 4
ET	Polar, tundra	0 < Th < 10

MAP, mean annual precipitation; MAT, mean annual temperature; Th, temperature of the hottest month; Tc, temperature of the coldest month; T10, number of months where the temperature is above 10; Pd, precipitation of the driest month; Pt (threshold) varies according to the following rules: if 70% of MAP occurs in winter then Pt = 2 x MAT, if 70% of MAP occurs in summer then Pt = 2 x MAT +28, otherwise Pt = 2 x MAT +14. Temperature in °C and precipitation in mm to withstand such changes, as evidenced by its ability to thrive under diverse precipitation patterns observed today (Fig. 4.10). While it could be hypothesized that *Rhizophora* might have outperformed *Pelliciera* during the EOT, challenging the idea that *Rhizophora* mangroves served as a refuge for *Pelliciera* survival and expansion, it is important to recognize that ecological competition does not necessarily lead to extinction. The coexistence of competing species is possible through niche segregation, a crucial element in maintaining biodiversity (MacArthur and Levins 1967; Violle et al. 2011; Kosicki 2022). Niche segregation can occur through spatial, temporal, and/or functional differences, allowing for the cohabitation of species with varying degrees of overlap. In the context of EOT mangroves, *Rhizophora* and *Pelliciera* likely differentiated in aspects such as dominance or autecological requirements (functional segregation) while sharing the same habitat (spatial overlapping).

Fernández-Palacios et al. (2021) investigated vegetation on oceanic islands and found that the primary species in major plant communities, termed "ecological winners," are usually immigrant generalist species. On the other hand, species successfully adapting to very specific environments through in situ evolution, known as "evolutionary winners," tend to thrive in marginal habitats and are labeled "ecological losers." In the post-Eocene Caribbean mangrove communities, a similar pattern is observed. *Rhizophora,* likely arriving from the AEP region through long-distance dispersal, is identified as an ecological winner. In contrast, *Pelliciera,* evolving in situ but eventually overshadowed by *Rhizophora* as the dominant species, is considered an evolutionary winner but an ecological loser. However, this situation is more complex because the ecological winner, *Rhizophora,* played a crucial role in creating microhabitat conditions that allowed the ecological loser, *Pelliciera,* to become an evolutionary winner. Despite losing dominance to *Rhizophora, Pelliciera* not only gained a more sheltered microhabitat from environmental stressors but also the ability to expand its geographical range, which was previously nonexistent. In ecological terms, competition seems to have led to mutualistic and facilitative relationships, mediated by functional niche segregation with some degree of physical overlap. See Chap. 6 for a more in-depth discussion of the ecological and evolutionary implications of the *Rhizophora–Pelliciera* interactions.

In addition to the rapid cooling process, the decrease in sea level during the EOT might have contributed to the mangrove turnover. As discussed in Sects. 4.1.1 and 4.2.3, a proposed global and regional (Caribbean) drop in sea level, estimated at around 70 meters (50 m below the current level), could have exposed shallow marine shelf environments. Apart from the examination of GAARlandia's presence or absence in Sects. 4.3.2 and 4.3.4, this decline in sea level may have enlarged and interconnected coastal areas conducive to the establishment and dispersal of mangroves (Fig. 4.4). This situation could have facilitated the documented expansion of mangroves during the Oligocene (Fig. 4.7), promoting the spread of euryoic and vagile mangrove elements like *Rhizophora*, which possess high dispersal capabilities, while potentially diminishing the prevalence of more specialized and less mobile species such as *Pelliciera*. Consequently, this dynamic could have exacerbated the impact of temperature decline, contributing to the mangrove revolution during the EOT.

The formation of the Panama isthmus during the Pliocene, by 3 Ma, disrupted the connection between the Pacific and Atlantic Oceans, causing changes in marine circulation patterns (O'Dea et al. 2016). Evaluating the effects of this event on mangroves is challenging due to limited Pliocene records, with only five available for comparison and just one providing quantitative data (Table 4.2). Caribbean mangroves that emerged after the Eocene directly preceded the mangroves existing today. The evolutionary process during the Neogene, as elaborated in the forthcoming chapter, was characterized by gradual diversification, with no evident extinctions or significant shifts in distribution ranges (Graham 1995; Rull 2023).

4.4 Synthesis

The EOT was characterized by rapid (0.5–0.8 my duration) worldwide cooling that caused the glaciation of Antarctica and represented the end of Cenozoic greenhouse climates and the onset of the present icehouse state of Earth. Whether this cooling was accompanied by a global trend toward aridification remains unclear, although the available evidence suggests this possibility at a regional level. A global sea-level fall of ~70 m (50 m below the present), on average, has been estimated for the EOT, which has been linked to the Antarctic glaciation. EOT cooling and Antarctic glaciation have been attributed to the interplay of three main mechanisms: a global reduction in atmospheric CO_2 concentration (from ~1000 ppm to 700–800 ppm), the establishment of the Antarctic Circumpolar Current (after the opening of the Drake Passage and the Tasman Gateway), and the albedo effect of ice sheet growth.

The EOT was marked by significant changes in biodiversity, driven by cooling and a decline in sea levels, along with reorganizations in paleogeography. Marine organisms, including foraminifers, radiolarians, calcareous nanoplankton, dinoflagellates, diatoms, and ostracods, experienced the most pronounced extinctions globally. On the other hand, terrestrial extinctions, particularly among plants and mammals, exhibited notable spatial variability. These extinctions during the Eocene, coupled with increased speciation in the Oligocene, set the stage for the development of modern biotas. Fossil evidence, primarily from pollen, suggests that the global disruption during the EOT played a crucial role in the divergence of mangrove biotas between the IWP and AEP regions. This disruption also contributed to the shift from *Pelliciera*-dominated to *Rhizophora*-dominated mangroves in the Neotropical region. Molecular phylogeographical studies propose that *Rhizophora* originated in the IWP during the Eocene and later migrated to the AEP region, achieving its current global distribution by the Late Eocene. Modern AEP *Rhizophora* species are believed to have emerged between the Late Miocene and the Pliocene. These studies emphasize that neither vicarianist nor dispersalist explanations alone can fully explain mangrove biogeography; both factors are essential for understanding the current patterns of these coastal ecosystems. At the E/O boundary, the Caribbean microplate, positioned between the North American and South

American plates, was moving eastward. The advancing front was situated between the present eastern Greater Antilles in the north and the Aves Ridge in the south, while the Lesser Antilles had not yet formed.

The majority of paleogeographical reconstructions indicate that during the EOT, the Caribbean region served as an open seaway linking the Pacific and Atlantic Oceans. This connectivity resulted from the presence of volcanic island arcs forming the eastern and western plate boundaries, facilitating marine circulation. Nonetheless, there are alternative theories proposing the existence of a temporary land bridge named GAARlandia, lasting less than three million years. This bridge would have connected the Greater Antilles with northern South America, formed by a combination of decreased sea levels and tectonic uplift. If GAARlandia indeed existed, it would have obstructed the marine connection between the Atlantic and Pacific Oceans. Paleoenvironmental records across the Circum-Caribbean region suggest that the EOT lasted between 33.8 and 33.5 Ma and unfolded in three distinct stages. This transitional period was marked by an overall cooling of 3–6 °C and a substantial eustatic sea-level decline of 67 m.

The EOT holds significant importance in shaping current biotic patterns in the Caribbean. While some scholars argue that the primary source of Caribbean biota is long-distance dispersal through marine currents from Central and South America, others contend that terrestrial migration from South America via GAARlandia played a crucial role. Surprisingly, the potential impact of EOT cooling and sea-level drops is seldom discussed. Most biogeographical studies in the Caribbean focus on the origins of island biotas, particularly in the Greater Antilles, with an emphasis on continental dispersal pathways. However, mangroves grow on continental and island coasts, and understanding their biogeographical patterns and processes is a more complex task. A comprehensive review of over 80 fossil pollen records, spanning from the Eocene to the Pliocene, with a focus on the EOT, has been conducted.

The dominant mangrove-forming tree in the Eocene Caribbean mangroves was the autochthonous *Pelliciera*, contributing up to approximately 60% of the pollen, alongside the intertidal palm *Nypa* and other back-mangrove elements like the fern *Acrostichum* in brackish waters and the freshwater palm *Mauritia*. *Rhizophora* was notably scarce, suggesting a potential Late Eocene introduction from the IWP through long-distance dispersal across the Atlantic. This idea is supported by the high dispersal capacity of *Rhizophora* propagules and ocean current modeling on a global scale. According to the GAARlandia hypothesis, *Pelliciera* mangroves would have been geographically isolated from Atlantic influence, connecting solely to Pacific coasts. This scenario would hinder the arrival of *Rhizophora* from the Atlantic, as the GAARlandia barrier acted as an obstacle. However, it is important to note that this land bridge was temporary, and for the majority of the Eocene, there was a connection to the Atlantic. Consequently, the influence of GAARlandia on the distribution of mangroves, particularly *Rhizophora*, seems to have been limited.

After the Eocene, *Rhizophora* became the dominant species in mangrove ecosystems, overshadowing *Pelliciera*, which was a minor component with pollen concentrations below 10% and was frequently absent. Consequently, the EOT marked a widespread and sudden substitution of *Pelliciera* by *Rhizophora* mangroves throughout the entire Caribbean region. This shift in dominance was accompanied by the expansion of mangroves from their original location in the Central America/NW South America sector (CA/NWSA) to encompass the entire circum-Caribbean region. The post-Eocene expansion of mangroves was primarily led by *Rhizophora*, known for the ability of its propagules to float for extended periods, remaining viable for a year or more. In contrast, *Pelliciera* propagules could only float for about a week, with a viability of a few months at most. Despite these differences, *Pelliciera* also expanded its range, likely facilitated by the presence of *Rhizophora*. The latter provided protection against both biotic and abiotic stressors, a phenomenon observed in contemporary mangrove ecosystems.

The shift in mangrove dominance from *Pelliciera* to the newly prevalent *Rhizophora* during the EOT might have been influenced by the recorded cooling. The ability of *Rhizophora* to grow under a broad range of temperature and moisture conditions (euryclimatic) allowed it to withstand the cooling, whereas *Pelliciera*, which is stenoclimatic and limited to warm and wet climates, could not endure the changing conditions. The tolerance of *Rhizophora* to diverse climates likely played a crucial role in its ability to persist during the cooling, while *Pelliciera* survived by inhabiting protected microhabitats within the *Rhizophora* canopy. This transition in mangrove dominance during the EOT can be seen as a case of ecological competition without outright exclusion. Instead, it promoted the coexistence of competing species through niche segregation, allowing both *Rhizophora* and *Pelliciera* to thrive in their respective habitats.

References

Aitken T, Mann P, Escalona A et al (2011) Evolution of the Grenada and Tobago basins and implications for arc migration. Mar Pet Geol 28:235–258

Ali JR (2012) Colonizing the Caribbean: is the GAARlandia land-bridge hypothesis gaining a foothold? J Biogeogr 39:431–433

Ali JR, Hedges SB (2021) Colonizing the Caribbean: new geological data and an updated land-vertebrate colonization record challenge the GAARlandia land-bridge hypothesis. J Biogeogr 48:2699–2707

Allen RW, Collier JS, Stewart AG et al (2019) The role of arc migration in the development of the Lesser Antilles: a new tectonic model for the Cenozoic evolution of the eastern Caribbean. Geology 47:891–895

Alonso R, Crawford AJ, Bermingham E (2012) Molecular phylogeny of an endemic radiation of Cuban toads (Bufonidae: Peltophryne) based on mitochondrial and nuclear genes. J Biogeogr 39:434–451

Amador-Castro F, García-Cayuela T, Alper HS et al (2021) Valorization of pelagic *Sargassum* biomass into sustainable applications: current trends and challenges. J Environ Manag 283:112013

Anderberg AA, Rydin C, Källersjö M (2002) Phylogenetic relationships in the order Ericales s.l.: analyses of molecular data from five genes from the plastid and mithochondrial genomes. Am J Bot 89:677–687

Antonioli L, de Araújo V, Dino R (2015) Palynology of carcinolites and limestones from the Baunilha Grande Ecofacies of the Pirabas Formation (Miocene of Pará state, northeastern Brazil). J S Am Earth Sci 62:134–147

Bacon CD, Baker WJ, Simmons MP (2012) Miocene dispersal drives island radiations in the palm tribe Trachycarpeae (Arecaceae). Syst Biol 61:426–442

Baldauf G (1992) Middle Eocene through early Miocene diatom floral turnover. In: Prothero DR, Berggren WA (eds) Eocene-oligocene climatic and biotic evolution. Princeton University Press, Princeton, pp 310–326

Barker PF, Burrell J (1977) The opening of Drake Passage. Mar Geol 25:15–34

Barker PF, Thomas E (2004) Origin, signature and palaeoclimatic influence of the Antarctic Circumpolar Current. Earth-Sci Rev 66:143–162

Belanger PE, Matthews RK (1984) The foraminiferal isotopic record across the Eocene/Oligocene boundary at Deep Sea Drilling Project Site 540. Init Rep Deep Sea Drilling Project 77:589–592

Berggren WA, Kent DW, Swisher ICC et al (1995) A revised Cenozoic geochronology and chronostratigraphy. In: Berggren AW, Kent DV, Handerbol J (eds) Geochronology, time scales and global stratigraphic correlation: a unified temporal framework for an historical geology. Soc Sedim Geol Spec Publ, pp 129–212

Bierman PR, Shakun JD, Corbett LB et al (2016) A persistent and dynamic East Greenland Ice Sheet over the past 7.5 million years. Nature 540:256

Bijl PK, Schouten S, Sluijs A et al (2009) Early Palaeogene temperature evolution of the southwest Pacific Ocean. Nature 461:776–779

Boersma A, Premoli Silva I (1986) Terminal Eocene events – planktonic foraminifera and isotopic evidence. In: Pomerol C, Premoli Silva I (eds) Terminal eocene events. Elsevier, Amsterdam, pp 213–224

Bohaty SM, Zachos JC, Delaney ML (2012) Foraminiferal Mg/Ca evidence for Southern Ocean cooling across the Eocene–Oligocene transition. Earth Planet Sci Lett 317–318:251–261

Borhidi A (1985) Phytogeographic survey of Cuba. I. Thephytogeographic characteristics and evolution of the flora of Cuba. Acta Bot Hung 31:3–34

Borrelli C, Cramer BS, Katz ME (2014) Bipolar Atlantic deepwater circulation in the middle-late Eocene: effects of Southern Ocean gateway openings. Paleoceanography 29:308–327

Boucher DH, James S, Keeler KH (1982) The ecology of mutualism. Annu Rev Ecol Syst 13:315–347

Bremer B, Bremer K, Heidari N et al (2002) Phylogenetics of asterids based on 3 coding and 3 non-coding chloroplast DNA markers and the utility on non-coding DNA at higher taxonomic levels. Mol Phylogenet Evol 24:274–301

Bruno JF, Stachowicz JJ, Bertness MD (2003) Inclusion of facilitation into ecological theory. Trees 18:119–125

Callaway RM (1995) Positive interactions among plants (interpreting botamical progress). Bot Rev 61:306–349

Castillo-Cárdenas MF, Díaz-Gonzales F, Cerón-Souza I et al (2014) Jumping a geographical barrier: diversification of the mangrove species *Pelliciera rhizophorae* (Tetrameristaceae) across the Central American Isthmus. Tree Genet Genomes 11:1–11

Castillo-Cárdenas MF, Ramirez-Silva JA, Sanjur O (2015) Evidence of incipient speciation in the Neotropical mangrove *Pelliciera rhizophorae* (Tetrameristaceae) as revealed by molecular, morphological, physiological and climatic characteristics. Bot J Linn Soc 179:499–510

Châteauneuf J-J (1980) Palynostratigraphie et paléoclimatologie de l'Éocène Supérieur et de l'Oligocène du Bassin de Paris. Mém Bur Rech Géol Min 116:1–360

Cohen KM, Finney SC, Gibbard PL et al (2013) The ICS international chronostratigraphic chart. Episodes 36:199–204. (version 2020/10; regularly updated at www.stratigraphy.org/ICSChart/)

Collinson ME (1992) Vegetation and floristic changes around the Eocene-Oligocene boundary in western and central Europe. In: Prothero DR, Berggren WA (eds) Eocene-oligocene climatic and biotic evolution. Princeton University Press, Princeton, pp 437–450

Cornée J-J, BouDagher-Fadel M, Philippon M et al (2020) Paleogene carbonate systems of Saint Barthélemy, Lesser Antilles: stratigraphy and general organization. Newsl Stratigr 53:461–478

Cornée J-J, Münch P, Philippon M et al (2021) Lost islands in the northern Lesser Antilles: possible milestones in the Cenozoic dispersal of terrestrial organisms between South-America and the Greater Antilles. Earth-Sci Rev 217:103617

Coxall HK, Pearson PN (2007) The Eocene-Oligocene transition. In: Williams M, Haywood AM, Gregory FJ et al (eds) Deep-time perspectives on climate change: marrrying the signal from computer models and biological proxies. Geological Soc, London, pp 351–387

Coxall HK, Wilson PA (2011) Early Oligocene glaciation and productivity in the eastern equatorial Pacific: insights into global carbon cycling. Paleoceanography 26:PA2221

Coxall HK, Wilson PA, Palike H et al (2005) Rapid stepwise onset of Antarctic glaciation and deeper calcite compensation in the Pacific Ocean. Nature 433:53–57

Coxall HK, Huck CE, Huber M et al (2018) Export of nutrient rich Northern Component Water preceded early Oligocene Antarctic glaciation. Nat Geosci 11:190–196

Dangremond EM, Feller IC (2014) Functional traits and nutrient limitation in the rare mangrove *Pelliciera rhizophorae*. Aquat Bot 116:1–7

Dangremond EM, Feller IC, Sousa WP (2015) Environmental tolerances of rare and common mangroves along light and salinity gradients. Oecologia 179:1187–1198

Darlington PJJ (1983) The origin of the fauna of the Greater Antilles, with discussion of dispersal of animals over water and through the air. Q Rev Biol 13:274–300

Dávalos L (2004) Phylogeny and biogeography of Caribbean mammals. Biol J Linn Soc 81:373–394

De Lira Mota MA, Harrington G, Jones TD (2020) Organic-walled dinoflagellate cyst biostratigraphy of the upper Eocene to lower Oligocene Yazoo Formation, US Gulf Coast. J Micorpaleontol 39:1–26

De Schepper S, Gibbard PL, Salzmann U et al (2014) A global synthesis of the marine and terrestrial evidence for glaciation during the Pliocene Epoch. Earth-Sci Rev 135:83–102

DeConto RM, Pollard D (2003) Rapid Cenozoic glaciation of Antarctica induced by declining atmospheric CO_2. Nature 421:245–249

Deler-Hernandez A, Sykora V, Seidel M et al (2018) Multiple origins of the *Phaenonotum* beetles in the Greater Antilles (Coleoptera: Hydrophilidae): phylogeny, biogeography and systematics. Zool J Linnean Soc 183:97–120

Diester-Haass L, Zahn R (1996) Eocene-Oligocene transition in the Southern Ocean: history of water mass circulation and biological productivity. Geology 24:163–166

Dockeri DT, Lozouet P (2003) Biotic patterns in Eocene-Oligocene mollusks of the Atlantic Coastal Plain. In: Prothero DR, Ivany L, Nesbitt EA (eds) From Greenhouse to Icehouse: the Marine Eocene Oligocene Transition. Columbia University Press, New York, pp 303–340

Dodd RS, Afzal-Rafii ZA (2002) Evolutionary genetics of mangroves: continental drift to recent climate change. Trees 16:80–86

Dodd RS, Afzal-Rafii Z, Kashani N et al (2002) Land barriers and open oceans: effects on gene diversity and population structure in *Avicennia germinans* L. (Avicenniaceae). Mol Ecol 11:1327–1338

Duke NC (2017) Mangrove floristics and biogeography revisited: further deductions from biodiversity hot spots, ancestral discontinuities, and common evolutionary processes. In: Rivera-Monroy VH et al (eds) Mangrove ecosystems: a global biogeographic perspective. Springer, Berlin, pp 17–53

Duke NC (2020) A systematic revision of the vulnerable mangrove genus *Pelliciera* (Tetrameristaceae) in equatorial Ammerica. Blumea 65:107–120

Duke NC, Lo EYL, Sun M (2002) Global distribution and genetic discontinuities of mangroves – emergning patterns in the evolution of *Rhizophora*. Trees 16:65–79

Duke NC, Meyneccke J-O, Dittman S et al (2017) A world without mangroves? Science 317:41b–42b

Ehlers J, Gibbard PL (2007) The extent and chronology of Cenozoic Global Glaciation. Quat Int 164–165:6–20

Eldrett JS, Greenwood DR, Harding IC (2009) Increased seasonality through the Eocene to Oligocene transition in northern high latitudes. Nature 459:969–973

Ellison AM, Farnsworth EJ, Merkt RE (1999) Origins of mangrove ecosystems and the marine biodiversity anomaly. Glob Ecol Biogeogr 8:95–115

Elsik WC (1974) Characteristic Eocene plynomorphs in the Gulf Coast, USA. Palaeontographica B 146:65–87

Fernández-Palacios JM, Otto R, Borregaard MK et al (2021) Evolutionary winners are ecological losers among oceanic island plants. J Biogeogr 48:2186–2198

Fonseca LHM (2021) Combining molecular and geographical data to infer the phylogeny of Lamiales and its dispersal patterns in and out of the tropics. Mol Phylogenet Evol 164:107287

Francis JE, Poole I (2002) Cretaceous and Tertiary climates of Antarctica: evidence from fossil wood. Palaeogeogr Palaeoclimatol Palaeoecol 182:47–64

Frederiksen NO (1985) Review of early Tertiary sporomorph paleoecology. Am Assoc Strat Palynol Contr Ser 19:1–92

Fritsch PW (2003) Multiple geographic origins of Antillean *Styrax*. Syst Bot 28:421–430

Fuchs HP (1970) Ecological and palynological notes on *Pelliciera rhizophorae*. Acta Bot Neerl 19:884–894

Funakawa S, Nishi H, Moore TC et al (2006) Radiolarian faunal turnover and paleoceanographic change around the Eocene-Oligocene boundary in the Central Equatorial Pacific, ODP Leg 199, Holes 1218A, 1219A and 1220A. Palaeogeogr Palaeoclimatol Palaeoecol 230:183–203

Galeotti S, Bijl P, Brinkhuis H et al (2022) The Eocene-Oligocene boundary climate transition: an Antarctic perspective. In: Florindo F, Siegert M, De Santis L et al (eds) Antarctic climate evolution. Elsevier, Amsterdam, pp 297–361

Garrocq C, Lallemand S, Marcaillou B et al (2021) Genetic relations between the Aves Ridge and the Grenada back-arc Basin, East Caribbean Sea. J Geophys Res Solid Earth 126:e2020JB020466

Germeraad JH, Hopping CA, Muller J (1968) Palynology of Tertiary sediments from tropical areas. Rev Palaeobot Palynol 6:189–348

Geuten K, Smets E, Schols P et al (2004) Conflicting phylogenies of balsaminoid families and the polytomy in Ericales: combining data in a Bayesian framework. Mol Phylogenet Evol 31:711–729

Graham A (1977) New records of *Pelliciera* (Theaceae/Pellicieraceae) in the Tertiary of the Caribbean. Biotropica 9:48–52

Graham A (1990) Late Tertiary microfossil flora from the Republic of Haiti. Am J Bot 77:911–926

Graham A (1995) Diversification of Gulf/Caribbean mangrove communities through Cenozoic time. Biotropica 27:20–27

Graham A (1999) Late cretaceous and cenozoic history of North American vegetation. Oxford University Press, New York

Graham A (2000) Palynofloras and terrestrial environments in the Eocene of the Caribbean basin. In: Sloan L, Scmitz B, Aubry MP et al (eds), Early paleogene warm climates and biosphere dynamics. GFF 122, p 64

Graham A (2006) Paleobotanical evidence and molecular data in reconstructing the historical phytogeography of Rhizophoraceae. Ann Missouri Bot Gard 93:325–334

Graham A, Cozadd D, Areces-Mallea A et al (2000) Studies in Neotropical paleobotany. XIV. A palynoflora from the Middle Eocene Saramaguacán Formation of Cuba. Am J Bot 87:1526–1539

Greenwood DR, Wing SL (1995) Eocene continental climates and latitudinal temperature gradients. Geology 23:1044–1048

Gruas-Cavagnetto C, Laurain M, Meyer R (1980) Un sol de mangrove fossilise dans les lignites du Soissonnais (Ypresien) a verzeney (Marne). Geobios 13:795–801

Hadly EA, Spaeth PA, Li C (2009) Niche conservatism above the species level. Proc Natl Acad Sci USA 106:19707–19714

He Q, Bertness MD, Altieri AH (2013) Global shifts towards positive species interactions with increasing environmental stress. Ecol Lett 16:695–706

Hedges SB, Hass CA, Maxson L (1992) Caribbean biogeography—molecular evidence for dispersal in West Indian terrestrial vertebrates. Proc Natl Acad Sci USA 89:1909–1913

Helland PE, Holmes MA (1997) Surface textural analysis of quartz sand grains from ODP Site 918 off the southeast coast of Greenland suggests glaciation of southern Greenland at 11 Ma. Palaeogeogr Palaeoclimatol Palaeoecol 135:109–121

Hooker JJ, Collinson ME, Sille NP (2004) Eocene–Oligocene mammalian faunal turnover in the Hampshire Basin, UK: calibration to the global time scale and the major cooling event. J Geol Soc Lond 161:161–172

Houben AJP, van Mourik CA, Montanari A et al (2012) The Eocene–Oligocene transition: Changes in sea level, temperature or both? Palaeogeogr Palaeoclimatol Palaeoecol 335–336:75–83

Huber M (2008) A Hotter Greenhouse? Science 321:353–354

Huber M, Caballero R (2011) The early Eocene equable climate problem revisited. Clim Past 7:603–633

Huber M, Sloan LC (2000) Climatic responses to tropical sea surfacetemperature changes on a "greenhouse" Earth. Paleoceanography 15:443–450

Hutchinson DK, Coxall HK, O'Regan M et al (2019) Arctic closure as a trigger for Atlantic overturning at the Eocene-Oligocene transition. Nat Commun 10:3797

Hutchinson DK, Coxall HK, Lunt DJ et al (2021) The Eocene-Oligocene transition: a review of marine and terrestrial proxy data, models and model-data comparisons. Clim Past 17:269–315

Iturralde-Vinent MA (2006) Meso-Cenozoic Caribbean paleogeography: implications for the historical biogeography of the region. Int Geol Rev 48:791–827

Iturralde-Vinent MA, MacPhee RDE (1999) Paleogeography of the Caribbean region: implications for Cenozoic biogeography. Bull Am Mus Nat Hist 238:1–95

Jablonski D, Chaloner WG (1994) Historical perspective: extinctions in the fossil record. Phil Trans R Soc Lond B 344:11–17

Jaramillo C, Dilcher DL (2001) Middle Paleogene palynology of Central Colombia, South America: a study of pollen and spores from tropical latitudes. Palaeontographica B 258:87–213

Jiménez JA (1984) A hypothesis to explain the reduced distribution of the mangrove *Pelliciera rhizophorae* Tr and Pl. Biotropica 16:304–308

Jin J-H, Herman AB, Spicer RA et al (2017) Palaeoclimate background of the diverse Eocene floras of South China. Sci Bull 62:1501–1503

Katz ME, Miller KG, Wright JD, Wade et al (2008) Stepwise transition fromthe Eocene greenhouse to the Oligocene icehouse. Nat Geosci 1:329

Keigwin LD, Croliss BH (1986) Stable isotopes in late Middle Eocene to Oligocene foraminifera. Geol Soc Am Bull 97:335–345

Keller G, Macleod N, Barrera E (1992) Eocene-Oligocene faunal turnover in planktonic foraminifera, and Antarctic glaciation. In: Prothero DR, Berggren WA (eds) Eocene-Oligocene climatic and biotic evolution. Princeton University Press, Princeton, pp 218–244

Kosicki JZ (2022) Niche segregation on the landscape scale of two co-existing related congeners in the sympatric zone – modelling approach. Ecol Model 468:109960

Kottek M, Grieser J, Beck C et al (2006) World map of the Köppen-Geiger climate classification updated. Meteorol Zeitschrift 15:259–263

Langton SJ, Rabideaux NM, Borrelli C et al (2016) Southeastern Atlantic deep-water evolution during the late-middle Eocene to earliest Oligocene (Ocean Drilling Program Site 1263 and Deep Sea Drilling Project Site 366). Geosphere 12:1032–1047

Larsen HC, Saunders AD, Clift PD et al (1994) Seven million years of glaciation in Greenland. Science 264:952–955

Liu Y-J, Arens NC, Li C-S (2007) Range change in *Metasequoia*: relationship to palaeoclimate. Bot J Linn Soc 154:115–127

Liu Z, Pagani M, Zinniker D et al (2009) Global cooling during the Eocene-Oligocene climate transition. Science 27:1187–1190

Livermore R, Hillenbrand C-D, Meredith M et al (2007) Drake Passage and Cenozoic climate: an open and shut case? Geochem Geophys Geosyst 8:Q01005

Lo EYY, Duke NC, Sun M (2014) Phylogeographic pattern of *Rhizophora* (Rhizophoraceae) reveals the importance of both vicariance and long-distance oceanic dispersal to modern mangrove distribution. BMC Evol Biol 14:83

Lorente MA (1986) Palynology and palynofacies of the Upper Tertiary in Venezuela. Dissert Bot 99:1–222

Losos JB, Schluter D (2000) Analysis of an evolutionary species-area relationship. Nature 408:847–850

MacArthur R, Levins R (1967) The limiting of similarity, convergence, and divergence of coexisting species. Am Nat 10:377–385

Machin J (1971) Plant microfossils from Tertiary deposits of the Isle of Wight. New Phytol 70:851–872

MacPhee RDE, Iturralde-Vinent M (1994) First Tertiary land mammal from Greater Antilles: an early miocene sloth (Xenarthra, Megalonychidae) from Cuba. Am Mus Nat Hist Novit 3094:1–13

MacPhee RDE, Iturralde-Vinent M (1995) Origin of the Greater Antillean land mammal fauna, 1: New Tertiary fossils from Cuba and Puerto Rico. Am Mus Nat Hist Novit 3141:1–31

MacPhee RDE, Iturralde-Vinent M (2000) A short history of Greater Antillean land mammals: biogeography, paleogeography, radiations, and extintions. Tropics 10:145–154

MacPhee RDE, Iturralde-Vinent M (2005) The interpretation of Caribbean paleogeography: reply to Hedges. Monogr Soc Hist Nat Balears 12:175–184

Mann P (2021) Gulf of Mexico, Central America, and the Caribbean. In: Alderton D, Elias SA (eds) Encyclopedia of geology. Academic Press, London, pp 47–67

Mann P, Taylor FW, Lawrence R et al (1995) Actively evolving microplate formation by oblique collision and sideways motion along strike-slip faults: an example from the northeastern Caribbean plate margin. Tectonophysics 246:1–69

Mann P, Rogers R, Gahagan L (2006) Overview of plate tectonic history and its unresolved tectonic problems. In: Bundschuh J, Alvarado GE (eds) Central America: geology, resources and hazards. Taylor and Francis/Balkema, Leiden, pp 201–237

Markwick PJ (2007) The palaeogeographic and palaeoclimatic significance of climate proxies for data-model comparisons. In: Williams M, Haywood AM, Gregory FJ et al (eds) Deep-time perspectives on climate change: marrying the signal from computer models and biological proxies. Geological Soc, London, pp 251–312

Maslin MA, Li XS, Loutre M-F et al (1998) The contribution of orbital forcing to the progressive intensification of Northern Hemisphere glaciation. Quat Sci Rev 17:411–426

Maurin O, Anest A, Bellot S et al (2021) A nuclear phylogenetic study of the angiosperm order Myrtales, exploring the potential and limitations of the universal Angiosperms353 probe set. Am J Bot 108:1087–1111

McCoy ED, Heck KL (1976) Biogeography of corals, seagrasses, and mangroves: an alternative to the center of origin concept. Syst Zool 25:201–210

Meng J, McKenna MC (1998) Faunal turnovers of Paleogene mammals from the Mongolian Plateau. Nature 394:364–357

Miller KG, Browning JV, Aubry M-P et al (2008) Eocene–Oligocene global climate and sea-level changes: St. Stephens Quarry, Alabama. Geol Soc Am Bull 120:34–53

Moen DS, Wiens J (2009) Phylogenetic evidence for competitively driven divergence: body-size evolution in Caribbean treefrogs (Hylidae: *Osteopilus*). Evolution 63:195–214

Moore MJ, Janzen RK (2005) Molecular evidence for the age, origin and evolutionary history of the American plant desrt genus *Tiquilia* (Boraginaceae). Mol Phylogenet Evol 39:668–687

Muellner-Riehl A, Rojas-Andrés BM (2022) Biogeography of Neotropical Meliaceae: geological connections, fossil and molecular evidence revisited. Brazil J Bot 45:527–543

Muller J, Di Giacomo E, Van Erve AW (1987) A palynological zonation for the cretaceous, tertiary and quaternary of northern South America. Am Assoc Strat Palynol Contr Ser 19:7–76

Nebelsick JH, Rasser MW, Bassi D (2005) Facies dynamics in Eocene to Oligocene circumalpine carbonates. Facies 51:197–216

Neill I, Kerr AC, Hastie AR et al (2013) The Albian-Turonian island arc rocks of Tobago, West Indies: geochemistry, petrogenesis, and Caribbean plate tectonics. J Petrol 54:1607–1639

O'Dea A, Lessios HA, Coates AG et al (2016) Formation of the Isthmus of Panama. Sci Adv 2:e1600883

Öğretmen N, Schiebel R, Jochum KP et al (2020) Deep thermohaline circulation across the closure of the Central American Seaway. Paleoceanogr Paleoclimatol 35:PA004049
Pagani M, Zachos JC, Freeman KH et al (2005) Marked decline in atmospheric carbon dioxide concentrations during Paleogene. Science 309:600–603
Pagani M, Huber M, Liu Z et al (2011) The role of carbon dioxide during the onset of Antarctic glaciation. Science 334:1261–1264
Pearson PN, McMillan IK, Wade BS et al (2008) Extinction and environmental change across the Eocene-Oligocene boundary in Tanzania. Geology 36:179–182
Pederneiras LC, Gaglioti AL, Romaniuc-Neto S et al (2018) The role of biogeographical barriers and bridges in determining divergent lineages in *Ficus* (Moraceae). Bot J Linn Soc 187:594–613
Peel MC, Finlayson BL, McMahon TA (2007) Updated world map of the Köppen-Geiger climate classification. Hydrol Earth-Syst Sci 11:1633–1644
Pennington RT, Dick CW (2004) The role of immigrants in the assembly of the South American rainforest tree flora. Philos Trans R Soc B 359:1611–1622
Pérez LF, Nielsen T, Knutz PC et al (2018) Large-scale evolution of the central-east Greenland margin: new insights to the North Atlantic glaciation history. Glob Planet Chang 163:141–157
Philippon M, Cornée J-J, Münch P et al (2020) Eocene intra-plate shortening responsible for the rise of a faunal pathway in the northeastern Caribbean realm. PLoS One 15:e0241000
Pindell JL (1990) Geological arguments suggesting a Pacific origin for the Caribbean Plate. In: Larue DK, Draper G (eds) Transactions of the 12th Caribbean Geologic Conference. Miami Geological Society, Miami, pp 1–4
Pindell J, Dewey JF (1982) Permo-Triassic reconstruction of western Pangea and the evolution of the Gulf of Mexico/Caribbean region. Tectonics 1:179–211
Pindell JL, Kennan L (2009) Tectonic evolution of the Gulf of Mexico, Caribbean and northern South America in the mantle reference frame: an update. Geol Soc London Spec Publ 328:1–55
Pindell JL, Cande SC, Pitman WC et al (1988) A plate-kinematic framework for models of Caribbean evolution. Tectonophysics 155:121–138
Pindell J, Kennan L, Stanek KP et al (2006) Foundations of Gulf of Mexico and Caribbean evolution: eight controversies resolved. Geol Acta 4:303–341
Plaziat J-C, Cavagnetto C, Koeniguer J-C et al (2001) History and biogeography of the mangrove ecosystem, based on a critical reassessment of the paleontological record. Wetl Ecol Manag 9:161–179
Pocknall DT, Erlich RN (2020) Palynostratigraphy and lithostratigraphy of Upper Cretaceous and Paleogene outcrop sections, Mérida Andes (Maracaibo Basin), Western Venezuela. J S Am Earth Sci 104:102830
Pocknall DT, Clowes CD, Jarzen DM (2022) *Spinizonocolpites prominatus* (McIntyre) Stover & Evans: fossil *Nypa* pollen, taxonomy, morphology, global distribution, and paleoenvironmental significance. N Zealand J Geol Geophys 66:558–570
Pound MJ, Salzmann U (2017) Heterogeneity in global vegetation and terrestrial climate change during the late Eocene to early Oligocene transition. Sci Rep 7:43386
Premoli Silva I, Jenkins DG (1993) Decision on the Eocene-Oligocene boundary stratotype. Episodes 16:379–382
Prothero D (2012) Cenozoic mammals and climate change: the contrast between coarse-scale versus high-resolution studies explained by species sorting. Geosciences 2:25–41
Prothero DR, Begggren WA (1992) Eocene-Oligocene climatic and biotic evolution. Princeton, Princeton University Press
Quattrocchio ME, Martínez MA, Hinojosa LF et al (2013) Quantitative analysis of Cenozoic palynofloras from Patagonia, southern South America. Palynology 37:246–258
Rabinowitz D (1978) Early growth of mangrove seedlings in Panama and a hypothesis concernign the relationship of dispersal and zonation. J Biogeogr 5:113–133
Raymo ME (1994) The initiation of Northern Hemisphere glaciation. Annu Rev Earth Planet Sci 22:353–383

Regalado L, Loriga J, Bechteler J et al (2018) Phylogenetic biogeography reveals the timing and-source areas of the *Adiantum* species (Pteridaceae) in the West Indies, with a special focus on Cuba. J Biogeogr 45:541–551

Romito S, Mann P (2020) Tectonic terrains underlying the present-day Caribbean plate: their tectonic origin, sedimentary thickness, subsidence histories and regional controls on hydrocarbon resources. In: Davidson I, Hull JNF, Pindell J (eds) The basins, orogens, and evolution of the Southern Gulf of Mexico and Northen Caribbean. Geol Soc, London, pp 343–378

Roncal J, Nieto-Blázquez ME, Cardona A et al (2020) Historical biogeography of Caribbean pnats revises regional paleoegeography. In: Rull V, Carnaval AC (eds) Neotropical diversification: patterns and processes. Springer Nature, Cham, pp 521–546

Rull V (1998a) Middle Eocene mangroves and vegetation changes in the Maracaibo basin, Venezuela. PALAIOS 13:287–296

Rull V (1998b) Evolución de los manglares neotropicales: la crisis del Eoceno. Interciencia 23:355–362

Rull V (2022a) The Caribbean mangroves: an Eocene innovation with no Cretaceous precursors. Earth-Sci Rev 231:104070

Rull V (2022b) Responses of Caribbean mangroves to Quaternary climatic, eustatic and anthropogenic drivers of ecological change: a review. Plan Theory 11:3502

Rull V (2023) The Neogene-Quaternary diversification trend in the shaping of modern Caribbean mangroves. Quat Sci Rev 300:107920

Saunders JB, Bernoulli D, Müller-Metz E et al (1984) Stratigraphy of the late Middle Eocene to Early Oligocene in the Bath Cliff section, Barbados, West Indies. Micropaleontology 30:390–425

Schellenberg SA (1998) Ecological and evolutionary response of deep-sea ostracodes to Eocene-Oligocene transition: high-resolution data from ODP site 744, Kerguelen Plateau. GSA Annual Meeting Abstracts, A-286

Schönenberger J, Anderberg AA, Sytsma KJ et al (2004) Molecular phylogenetics and patterns of floral evolution in the Ericales. Int J Plant Sci 166:265–288

Schwarzbach AE, Ricklefs RE (2000) Systematic affinities of Rhizophoraceae and Anisophylleaceae, and intergeneric relationships within Rhizophoraceae, based on chloroplast DNA, nuclear ribosomal DNA, and morphology. Am J Bot 87:547–564

Schwarzbach AE, McDade LA (2002) Phylogenetic relationships of the mangrove family Avicenniaceae based on chloroplast and nuclear ribosomal DNA sequences. Syst Bot 27:84–98

Sepkoski JJ (1986) Phanerozoic overview of mass extinction. In: Raup D, Jablonski D (eds) Patterns and processes in the history of life. Springer, Berlin, pp 277–295

Shackleton NJ, Backman J, Zimmerman H et al (1984) Oxygen isotope calibration of the onset of ice-rafting and history of glaciation in the North Atlantic region. Nature 307:620–623

Sheldon ND, Grimes ST, Hooker JJ et al (2016) Coupling of marine and continental oxygen isotope records during the Eocene-Oligocene transition. Geol Soc Am Bull 128:502–510

Sluiter IRK, Holdgate GR, Reichgelt T et al (2022) A new perspective on Late Eocene and Oligocene vegetation and paleoclimates of South-eastern Australia. Palaeogeogr Palaeoclimatol Palaeoecol 596:110985

Smith SA, Brown JW (2018) Constructing a broadly inclusive seed plant phylogeny. Am J Bot 105:302–314

Sowunmi MA (1986) Change of vegetation with time. In: Lawson GW (ed) Plant ecology in West Africa. Wiley, Chichester, pp 273–307

Srivastava J, Prasad V (2015) Effect of global warming on diversity pattern in *Nypa* mangroves across the Paleocene-Eocene transition in the paleo-equatorial region of the Indian subcontinent. Palaeogeogr Palaeoclimatol Palaeoecol 429:1–12

Srivastava J, Prasad V (2018) Evolution and paleobiogeography of mangroves. Mar Ecol 40:e12571

Stachowicz JJ (2001) Mutualism, facilitation, and the structure of ecological communities. Bioscience 51:235–246

Straume EO, Nummelin A, Gaina C et al (2022) Climate transition at the Eocene-Oligocene influenced by bathymetric changes to the Atlantic-Arctic ocean gateways. Proc Natl Acad Sci USA 119:e2115346119

Su T, Li S-F, Tang H et al (2018) *Hemitrapa* Miki (Lythraceae) from the earliest Oligocene of southeastern Qinghai-Tibetan Plateau and its phytogeographic implications. Rev Palaeobot Palynol 257:57–63

Sun J, Ni X, Bi S et al (2014) Synchronous turnover of flora, fauna and climate at the Eocene-Oligocene boundary in Asia. Sci Rep 4:7463

Takayama K, Tateishi Y, Kajita T (2021) Global phylogeography of a pantropical mangrove genus *Rhizophora*. Sci Rep 11:7228

Tomlinson PB (2016) The botany of Mangroves. Cambridge University Press, Cambridge

Tong Y, Binford G, Rheims CA et al (2019) Huntsmen of the Caribbean: multiple tests of the GAARlandia hypothesis. Mol Phylogenet Evol 130:259–268

Triest L (2008) Molecular ecology and biogeography of mangrove trees towards conceptual insights on gene flow and barriers: a review. Aquat Bot 89:138–154

Van der Hammen T, Wijmstra TA (1964) A palynological study on the tertiary and upper cretaceous of British Guiana. Leidse Geol Mededel 30:183–241

Van der Stocken T, Carroll D, Menemenlis D et al (2019) Global-scale dispersal and connectivity in mangroves. Proc Natl Acad Sci USA 116:915–922

Van Ee BW, Berry PE, Riina R et al (2008) Molecular phylogenetics and biogeography of the Caribbean-centered *Croton* subgenus *Moacroton* (Euphorbiaceae s.s.). Bot Rev 74:132–165

Van Mourik CA, Brinkhuis H (2005) The Massignano Eocene-Oligocene golden spike section revisited. Stratigraphy 2:13–29

Van Steenis CGGJ (1962) The distribution of mangrove plant genera and its significance for palaeogeography. Proc Kon Ned Akad Wetesnsch 65:164–169

Violle C, Nemergut DR, Pu Z et al (2011) Phylogenetic limiting similarity and competitive exclusion. Ecol Lett 14:782–787

Wade BS, Houben AJP, Quaijtaal W et al (2012) Multiproxy record of abrupt sea-surface cooling across the Eocene-Oligocene transition in the Gulf of Mexico. Geology 40:159–162

Westerhold T, Marwan N, Drury AJ et al (2020) An astronomically dated record of earth's climate and its predictability over the last 66 million years. Science 369:1383–1387

Wiens JJ, Graham CH (2005) Niche conservatism: integrating evolution, ecology, and conservation biology. Annu Rev Ecol Evol Syst 36:519–539

Wiens JJ, Ackerly DD, Allen AP et al (2010) Niche conservatism as an emerging principle in ecology and conservation biology. Ecol Lett 13:1310–1324

Wilson DS, Pollard D, DeConto RM et al (2013) Initiation of the West Antarctic Ice Sheet and estimates of total Antarctic ice volume in the earliest Oligocene. Geophys Res Lett 40:4305–4309

Woodburne MO, Goin FJ, Bond M et al (2014) Paleogene land mammal faunas of South America; a response to global climatic changes and indigenous floral diversity. J Mammal Evol 21:1–73

Xie X, Mann P, Escalona A (2010) Regional provenance study of Eocene clastic sedimentary rocks within the South America-Caribbean plate boundary zone using detrital zircon geochronology. Earth Planet Sci Lett 291:159–171

Xu S, He Z, Zhang Z et al (2017) The origin, diversification and adaptation of a major mangrove clade (Rhizophoraceae) revealed by whole-genome sequencing. Natl Sci Rev 4:721–734

Yancey TE, Elsik WC, Sancay RH (2003) The palynological record of late Eocene climate change, northwest Gulf of Mexico. In: Prothero DR, Ivany L, Nesbitt EA (eds) From greenhouse to icehouse: the Marine Eocene Oligocene transition. Columbia University Press, New York, pp 252–268

Young DA, Wright AP, Roberts JL et al (2011) A dynamic early East Antarctic Ice Sheet suggested by ice-covered fjord landscapes. Nature 474:72

Zachos JC, Quinn TM, Salamy KA (1996) High-resolution (104 years) deep-sea foraminiferal stable isotope records of the Eocene-Oligocene climate transition. Paleoceanography 11:251–266

Zachos J, Pagani M, Sloan L et al (2001) Trends, rhythms, and aberrations in global climate 65 Ma to present. Science 292:686–693

Neogene Diversification 5

Abstract

During the Oligocene, the Caribbean mangroves were dominated by *Rhizophora* alone, as the other mangrove-forming trees (*Avicennia*, *Laguncularia*) and most mangroves associates were still absent. The diversification of Caribbean mangroves occurred during the Neogene and was complete in the Quaternary, in a so-called Neogene–Quaternary diversification trend (NQDT). This trend was marked by a rise of 25 genera compared to the 7 representatives from the Paleogene. The Miocene saw the emergence of only two genera, with the most significant increases occurring during the transition from Pliocene to Quaternary and in the contemporary living record. Half of the true-mangrove genera (*Rhizophora*, *Pelliciera*, *Acrostichum*) were already in existence before the Neogene, while the others gradually appeared in the Oligo-Miocene (*Crenea*), Early–Middle Miocene (*Avicennia*), and Mio-Pliocene (*Laguncularia*). None of the current associate mangrove genera were present during the Paleogene; all of them arose in the Miocene (23 genera) or during the Oligo-Miocene transition (3 genera), playing a major role in the NQDT in terms of sheer numbers. There have been no documented regional extinctions at the generic level in Caribbean mangroves since the Miocene. These findings call for additional in-depth quantitative studies to explore potential causal relationships with climatic, eustatic, and paleogeographical changes.

In the preceding chapter, we explored the emergence of the precursors of modern Caribbean mangroves following the EOT global disruption. This transition marked a shift from Eocene *Pelliciera*-dominated mangroves to the dominance of *Rhizophora* in the post-Eocene. Notably, during the Oligocene, *Rhizophora* was the sole documented mangrove tree, with other extant genera such as *Avicennia* and

V. Rull, *Origin and Evolution of Caribbean Mangroves*, Ecological Studies 252,
https://doi.org/10.1007/978-3-031-57612-6_5

Laguncularia still absent, as occurred with the majority of other contemporary mangrove taxa (Table 2.2). Therefore, the diversification of Caribbean mangroves, which led to their current taxonomic composition, occurred after the Oligocene. This phenomenon is referred to as the Neogene–Quaternary diversification trend (NQDT). Graham (1995) was the first to acknowledge this trend (although he did not use the term NQDT), emphasizing that the diversification of the Caribbean/Gulf of Mexico mangroves occurred mainly in the Neogene, when the number of mangrove elements increased from three, in the Eocene, to the current 27 genera. Although Graham (1995) proposed the extinct *Brevitricolpites variabilis* as a potential fourth Eocene mangrove component, its taxonomic identity and environmental features were later questioned in subsequent studies (Jaramillo and Dilcher 2001). The original characterization of the NQDT was based on fossil pollen evidence from previously published range charts (Germeraad et al. 1968; Muller et al. 1987; Lorente 1986). This information was supplemented by additional records from the Miocene, Pliocene, and Quaternary periods (Graham 1995). In this work, we update the NQDT using the CARMA dataset (Table 2.4, Fig. 2.8), which incorporates over 110 pollen records from Neogene, Quaternary, and modern times (Fig. 5.1). This analysis specifically focuses on diversification trends, while Chap. 6 addresses Neogene biogeographical shifts, and Chap. 7 delves into mangrove responses to Pleistocene environmental changes and Holocene anthropization.

Among the currently existing mangrove species in the Caribbean (Table 2.2), only eight possess fossil counterparts or morphospecies (Table 5.1), and the remaining species are documented in fossil records using their current taxonomic names, typically at the genus level. This is primarily because distinguishing between species based solely on pollen morphology is unfeasible. Consequently, for the diversification analysis conducted in this study, extant genus names are utilized as the operational taxonomic unit to standardize the data. The focus of the analysis is on diversification, defined as the net increase in diversity, taking into account the balance between speciation and extinction (Cracraft 1985), and is expressed in terms of taxa richness.

5.1 Neogene

Figure 5.2 depicts the occurrence ranges of Caribbean genera with Neogene fossil pollen representatives, indicating the localities from which the evidence has been found (abbreviations in Table 5.1). Noteworthy, *Pelliciera*, *Rhizophora*, and *Acrostichum* first appeared prior to the Neogene, specifically in the Early Eocene, Middle Eocene, and Late Cretaceous, respectively, as discussed in Chap. 3. *Hibiscus* and *Crenea* made their first appearances in Oligocene and Oligo-Miocene sediments from Venezuela (Falcón; Fc) and Guyana (Shelter Belt; SB), respectively (Rull and Poumot 1997; Van der Hammen and Wijmstra 1964). *Pachira* and *Acacia* emerged for the first time during the Late Oligocene–Early Miocene in Venezuela (Maturín; Mn) and Colombia (Planeta Rica; Pn), respectively (Helenes and Cabrera 2002; Dueñas 1980). *Avicennia* first appeared in the Early/Middle Miocene in

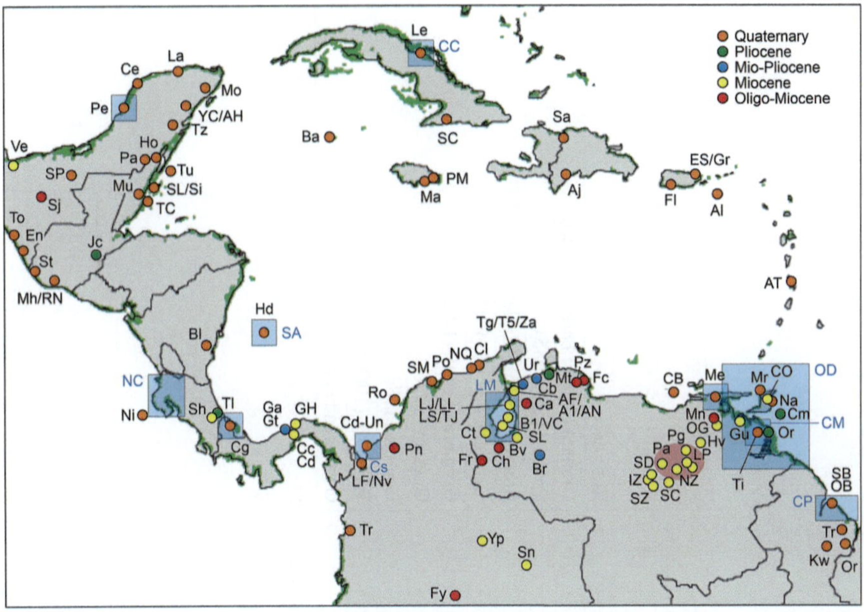

Fig. 5.1 Map of the Caribbean region indicating the localities with Neogene and Quaternary pollen records (colored dots) and the areas with modern-analog studies (brown boxes). The pink area in eastern Venezuela represents the cluster of sites where *Avicennia* showed its first appearances during the Early/Middle Miocene (see text for details). Locality codes according to Table 2.4 and Fig. 2.8

Venezuela (Los Pobres-1; LP) (Lorente 1986), while *Laguncularia* appeared for the first time in the Late Miocene in Mexico (Ve) (Graham 1976). These findings indicate that six out of the nine extant mangrove genera (considering *Acacia* as a mangrove element) with documented fossil pollen originated during the Neogene. The presence of *Avicennia* in the Early/Middle Miocene and *Laguncularia* in the Late Miocene suggests that all true Neotropical mangrove elements were already established in the Caribbean region by the end of the Neogene.

Due to the importance of *Avicennia* and *Laguncularia* as modern mangrove-forming trees (Chap. 2), their fossil records were analyzed in more depth to unravel their inception as relevant mangrove elements during the Neogene. The first Early/Middle Miocene *Avicennia* records (Fig. 5.2) consisted of single or scattered occurrences, barely distinguishable in standard pollen diagrams (Lorente 1986), located in a small spot of Venezuelan sites (NZ, Pa, Pg, LP) (Fig. 5.1), where *Rhizophora* largely dominated attaining pollen values up to 90% (Table 4.2). This indicates that *Avicennia* originated as a minor and localized mangrove element within *Rhizophora*-dominant mangroves and further disappeared from the fossil record until the Pliocene, when it reappeared in eastern Venezuela (Or) and northern Guyana (SB). The former is a qualitative record, but a percentage pollen diagram is available for the latter, showing that *Avicennia* was already an important component of the

Table 5.1 Neogene fossil pollen representatives of extant mangrove genera

Genera	Family	Fossil pollen	Reference
Acacia[a]	Fabaceae	*Acacia*/*Polyadopollenites mariae*	Graham (1976); Dueñas (1980)
Acrostichum	Pteridaceae	*Deltoidospora adriennis*	Frederiksen (1985)
Avicennia	Acanthaceae	*Retitricolpites* sp./*Avicennia*	Lorente (1986); Muller et al. (1987)
Crenea	Lythraceae	*Verrutricolporites rotundiporus*	Germeraad et al. (1968)
Hibiscus	Malvaceae	*Echiperiporites estelae*	Germeraad et al. (1968)
Laguncularia	Combretaceae	*Laguncularia*	Graham (1976)
Pachira	Bombacaceae	*Bombacacidites baculatus*	Lorente (1986)
Pelliciera	Pellicieraceae	*Psilatricolpoites crassus*	Wijmstra (1968)
Rhizophora	Rhizophoraceae	*Zonocostites ramonae*	Germeraad et al. (1968)

[a]Not included in Table 2.2 but considered to be a past mangrove element by some authors (Graham 1995)

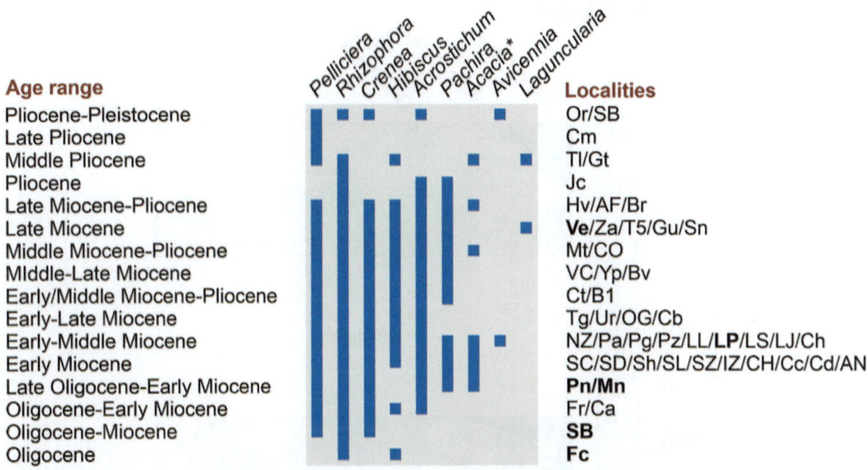

Fig. 5.2 Range chart of pollen from mangrove genera with Neogene fossil representatives, as indicated in the original references. Taxa not listed as mangrove elements (Table 2.2) but considered as such in the original references are marked with an asterisk. Sites recording the first appearances, as indicated in the text, are in bold. Site abbreviations, references, and other details in Table 2.4 and Figs. 2.8 and 5.1

Plio-Pleistocene mangrove communities (Fig. 5.3), similar to the present situation (Chap. 2). Therefore, with the available data, the modern relevance of *Avicennia* as an important mangrove-forming tree would not have been attained until the Plio-Pleistocene.

In contrast with *Avicennia*, the first appearance of *Laguncularia* (Late Miocene) occurred at relatively high abundances (up to 56%), suggesting that this tree was already an important component of *Rhizophora*-dominated mangroves (Table 4.2). The other possible *Laguncularia* record was from the Late Pliocene of Costa Rica (Talamanca) (Graham and Dilcher 1998) and was deduced from the presence of a pollen type named Unknown-3 (Graham and Dilcher 1998) and identified here as likely corresponding to *Laguncularia*. In summary, the available evidence suggests that *Laguncularia* and *Avicennia* began to be relevant mangrove elements in the Late Miocene and the Plio-Pleistocene, respectively. This means that the Caribbean, and likely the Neotropical, mangroves had been dominated by a single mangrove-forming tree, *Rhizophora*, for a large part of the Neogene, and the other major true-mangrove elements were not important until the last times of this geological period.

5.2 Quaternary

Quaternary records primarily rely on Holocene sequences since Pleistocene records are scarce, with only a few Late Pleistocene sections available (Rull 2022). The earliest documented occurrences trace back to the last glacial cycle (~130–70 cal kyr BP), when mangrove pollen assemblages were dominated by *Rhizophora*,

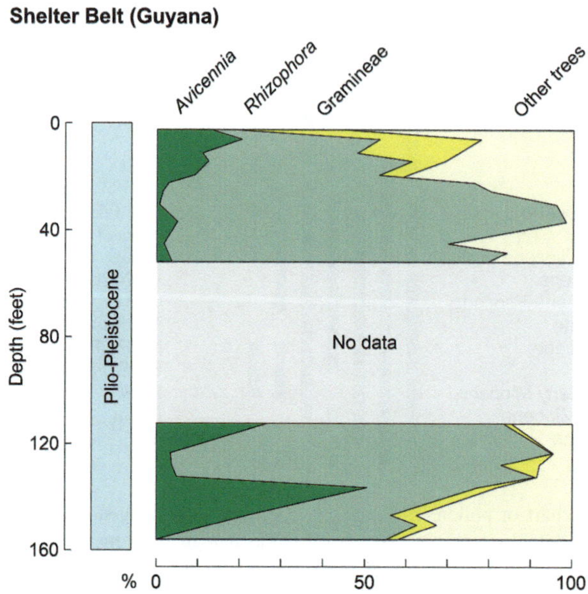

Fig. 5.3 Summary pollen diagram from the Plio-Pleistocene of the Shelter Belt borehole in Guyana. Redrawn and simplified from Van der Hammen and Wijmstra (1964)

Avicennia, and *Acrostichum* (Van der Hammen 1963; González and Dupont 2009; González et al. 2009). Following the Last Glaciation, Early Holocene sequences are largely absent in the Caribbean region, reappearing in the Middle Holocene (from 8.2 cal kyr BP onward) (Rull 2022). Based on available evidence, three genera (*Cassipourea*, *Pterocarpus*, and *Randia*) have been present in Central America (Gt; Panama) since the beginning of the Holocene (Bartlett and Barghoorn 1973) (Fig. 5.4). Three additional genera (*Ficus*, *Conocarpus*, and *Batis*) emerged in Cuba (La Leche; Le) and Mexico (Lake Tzib; Tz) at the onset of the Middle Holocene (Peros et al. 2007; Carrillo-Bastos et al. 2013). The initial occurrences of *Hippomane* and *Tabebuia* were documented in Trinidad (Maracas; Mr) and Venezuela (Tigre; Ti), respectively, during the Middle Holocene (Ramcharan and McAndrews 2006; Montoya et al. 2019). *Caesalpinia* also made its first appearance in Mexico (Rio Hondo; Ho), while *Dalbergia* was initially recorded in the Late Holocene of Colombia (Honda; Hd) (González et al. 2010; Aragón-Moreno et al. 2018). However, the absence of pollen records for much of the Pleistocene makes it impossible to ascertain whether these genera originated before the Holocene.

5.3 Modern

The existing research on contemporary pollen deposition contributes two additional genera to the list of sedimentary pollen. Specifically, *Rhabdadenia*, identified in the Orinoco delta and adjacent regions such as Playa Medina (Me) (Rull and

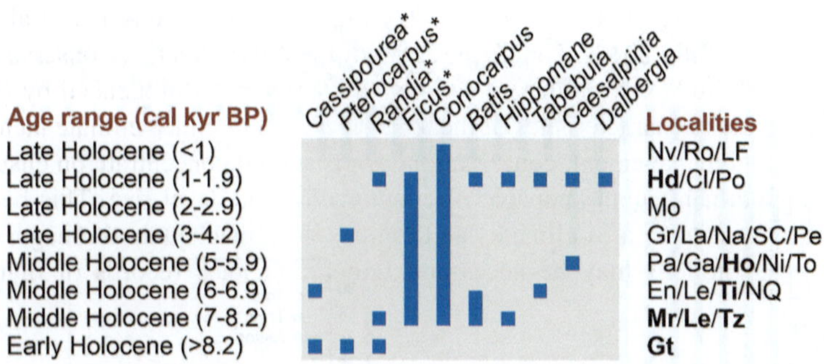

Fig. 5.4 Range chart of pollen types from mangrove genera with Holocene representatives. Taxa not listed as mangrove elements (Table 2.2) but considered as such in the original references are marked with an asterisk. Sites recording the first appearances, as indicated in the text, are in bold. Site abbreviations, references, and other details in Table 2.4 and Figs. 2.8 and 5.1

Vegas-Vilarrúbia 1999; Hofman 2002), and *Scaevola*, found in Cuba (Cayo Coco; CC) (Davidson 2007), are now included. The remaining genera listed in Table 2.2 (*Amoora*, *Amphitecna*, *Anemopaegna*, *Mora*, *Muellera*, *Pavonia*, *Phryganocydia*, *Pluchea*, *Rustia*, *Thespesia*, and *Tuberostylis*) are currently only recognized through their extant species.

5.4 Synthesis and Discussion

Figure 5.5 illustrates the results of the analysis in a graphical format, emphasizing the following key points:

- A total of 32 mangrove genera were identified in the Neogene, Quaternary, and modern records. Of them, six are classified as true mangroves, while the remaining 26 are either associates (21) or potential associates (5).
- The NQDT represented a significant increase (25 genera; 78% of the total) compared to the previously existing seven Paleogene representatives.
- Only two components (*Avicennia*, *Laguncularia*) appeared in the Miocene, with the most significant increases occurring during the PQ (Pliocene–Quaternary) transition (10 appearances) and the ML (modern-living) record (11 appearances).
- Three of the six true-mangrove elements (*Rhizophora*, *Pelliciera*, *Acrostichum*) emerged before the Neogene, while others (*Crenea*, *Avicennia*, *Laguncularia*) progressively appeared in the Oligo-Miocene (OM), Early–Middle Miocene, and Mio-Pliocene, respectively.
- Despite its first appearances in the Early/Middle Miocene, *Avicennia* did not become a significant mangrove component until the Plio-Pleistocene.
- All associate mangrove genera were absent in the Paleogene; they all emerged in the Miocene (23) and the OM transition (3), playing a crucial role in the NQDT.

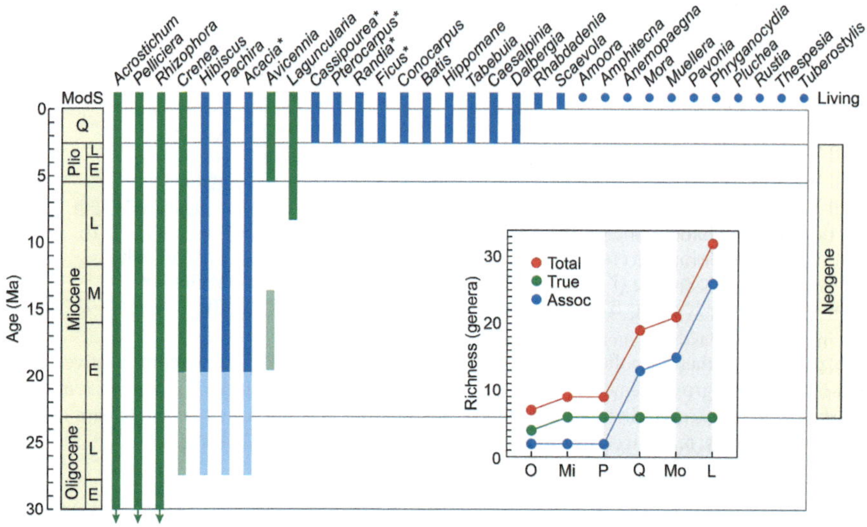

Fig. 5.5 Range chart of the Caribbean mangrove genera (blue dots represent living species with not known fossil representatives). True mangrove elements are in green and associate elements in blue (*elements considered to be mangrove associates in the original papers but not listed in Table 2.2). In the richness plot, gray bands represent acceleration phases. O, Oligocene; Mi, Miocene; P, Pliocene; Mo, modern; L, living

- No regional extinctions at the generic level have been recorded in Caribbean mangroves since the Miocene.

Certain points necessitate specific comments. The PQ (Pliocene–Quaternary) increase primarily occurred during the Pleistocene and Early Holocene, for which we lack comprehensive records, except for a few Late Pleistocene sequences. As a result, the observed acceleration in diversification mainly arises from comparing pollen records in Pliocene rocks and unconsolidated Mid–Late Holocene sediments. These records exhibit distinct taphonomic conditions, particularly concerning pollen preservation. To bridge this gap, it is crucial to investigate Pleistocene sediments, predominantly from marine cores like those retrieved in the Venezuelan Cariaco Basin (CB), which house extensive Pleistocene sequences (Haug et al. 1998; Yarincik and Murray 2000). Regarding the ML (modern-living) diversity increase, potential methodological artifacts cannot be dismissed. This is due to contemporary records relying on pollen from surface sediments, whereas living records relate to the presence of parent plants in the current flora. Comparing these records is challenging because the living plant record is likely more comprehensive, while the modern sedimentary record may be less complete and influenced by various factors, including pollen production, dispersal, sedimentation, preservation, reworking, and other taphonomic processes. Ideally, comparisons of fossil pollen records should be done with modern sedimentary pollen rather than with living taxa.

Additional relevant observations are that the primary structural components of mangroves exhibit significantly lower diversity than associated taxa. Their current richness appears to have been achieved in the Late Miocene–Pliocene, with *Avicennia* and *Laguncularia* emerging as significant elements. This implies that Neogene mangroves were predominantly structured around *Rhizophora*, serving as the singular dominant mangrove-forming tree. *Pelliciera*, once a prominent species, underwent substantial reduction and confined itself to specialized microhabitats by the end of the EOT, as discussed in Chap. 4. According to Duke (2017), it is likely that *Avicennia*, along with other globally distributed mangrove taxa, originated from a common source in the IWP region and followed dispersal patterns similar to *Rhizophora* through the Atlantic pathway to reach the Neotropics. This aligns with the widespread prevalence of *Avicennia*-dominated mangroves in Euro-African tropical areas between the Middle Miocene and the Pliocene (Popescu et al. 2021), crucial periods for the introduction and expansion of this taxon in the Caribbean region. Although *Laguncularia* has received less attention, its current distribution along the eastern American and western African coasts (Fig. 2.5) suggests that the Atlantic pathway may also have played a significant role in its evolutionary history. To validate these hypotheses, comprehensive phylogeographical and paleontological studies are needed, specifically focusing on the origin of these mangrove taxa in the Caribbean/Neotropical region.

The lack of extinctions and significant community turnover highlights the high resilience of Neogene Caribbean mangroves, despite the occurrence of significant changes in climate, eustatic conditions, and paleogeography (Westerhold et al. 2020). While the connection between Neogene mangrove diversification and external environmental shifts remains speculative due to the limited chronological resolution and scarcity of quantitative pollen records, it is crucial to highlight that the major diversification increase known as PQ acceleration occurred during the Pleistocene. This period was marked by frequent climatic and eustatic fluctuations associated with glacial–interglacial cycles. Although establishing causal relationships is premature, the alignment of diversification with Pleistocene conditions warrants further investigation and dedicated studies to explore potential causal connections.

The diversification trends identified in this analysis agree with earlier findings by Graham (1995) and contribute supplementary information through an extensive dataset covering the entire Caribbean region, both geographically and chronologically. Apart from offering insights into mangrove evolution, these pollen records serve as valuable calibration points, enhancing the accuracy of chronological interpretations in molecular phylogenetic and phylogeographical investigations of mangrove taxa (Graham 2006; Triest 2008; Lo et al. 2014; Xu et al. 2017; Takayama et al. 2021).

References

Aragón-Moreno AA, Islebe GA, Roy PD et al (2018) Climate forcings on vegetation of the southern Yucatán Peninsula (Mexico) during the middle to late Holocene. Palaeogeogr Palaeoclimatol Palaeoecol 495:214–226

Bartlett AS, Barghoorn ES (1973) Phytogeographic history of the Isthmus of Panama during the past 12,000 years (a history of vegetation, climate and sea-level change). In: Graham A (ed) Vegetation and vegetational history of Northern Latin America. Elsevier, Amsterdam, pp 203–299

Carrillo-Bastos A, Islebe GA, Torrescano-Valle N (2013) 3800 years of quantitative precipitation reconstruction from the northwest Yucatan Peninsula. PLoS One 8:e84333

Cracraft J (1985) Biological diversification and its causes. Ann Missouri Bot Gard 72:794–822

Davidson DE (2007) Modern pollen spectra from mangrove ecosystems of the sabana-camagüey archipelago and Ciego de Ávila, Cuba. MS diss, Univ Toronto, Canada

Dueñas H (1980) Palynology of Oligocene-Miocene strata of borehole Q-E-22, Planeta Rica, northern Colombia. Rev Palaeobot Palynol 30:313–328

Duke NC, Meyneccke J-O, Dittman S et al (2017) A world without mangroves? Science 317:41b–42b

Frederiksen NO (1985) Review of early Tertiary sporomorph paleoecology. Am Assoc Strat Palynol Contr Ser 19:1–92

Germeraad JH, Hopping CA, Muller J (1968) Palynology of Tertiary sediments from tropical areas. Rev Palaeobot Palynol 6:189–348

González C, Dupont LM (2009) Tropical salt marsh succession as sea-level indicator during Heinrich events. Quat Sci Rev 28:939–946

González C, Dupont LM, Behling H et al (2009) Neotropical vegetation response to rapid climate changes duing the last glacial period: palynological evidence from the Cariaco Basin. Quat Res 69:217–230

González C, Urrego LE, Martínez JI et al (2010) Mangrove dynamics in the southwestern Caribbean since the 'Little Ice Age': a history of human and natural disturbances. The Holocene 20:849–861

Graham A (1976) Studies in Neotropical paleobotany. II. The Miocene communties of Veracruz. Mexico Ann Missouri Bot Gard 63:787–842

Graham A (1995) Diversification of Gulf/Caribbean mangrove communities through Cenozoic time. Biotropica 27:20–27

Graham A (2006) Paleobotanical evidence and molecular data in reconstructing the historical phytogeography of Rhizophoraceae. Ann Missouri Bot Gard 93:325–334

Graham A, Dilcher DL (1998) Studies in Neotropical botany. XII. A palynoflora from the Pliocene Rio Banano Formation of Costa Rica and the Neogene vegetation of Mesoamerica. Am J Bot 85:1426–1438

Haug GH, Pedersen TF, Sigman DM et al (1998) Glacial/Interglacial variations in production and nitrogen fixation in the Cariaco Basin during the last 580 kyr. Paleoceanography 13:427–432

Helenes J, Cabrera D (2002) Oligo-Miocene palynomorph assemblages from eastern Venezuela. Palynology 27:5–25

Hofman C-C (2002) Pollen distribution in sub-Recent sedimentary environments of the Orinoco Delta (Venezuela) – an actuo-palaeobotanical study. Rev Palaeobot Palynol 119:191–217

Jaramillo C, Dilcher DL (2001) Middle Paleogene palynology of Central Colombia, South America: a study of pollen and spores from tropical latitudes. Palaeontographica B 258:87–213

Lo EYY, Duke NC, Sun M (2014) Phylogeographic pattern of *Rhizophora* (Rhizophoraceae) reveals the importance of both vicariance and long-distance oceanic dispersal to modern mangrove distribution. BMC Evol Biol 14:83

Lorente MA (1986) Palynology and palynofacies of the Upper Tertiary in Venezuela. Dissert Bot 99:1–222

Montoya E, Pedra-Méndez J, García-Falcó E et al (2019) Long-term vegetation dynamics of a tropical megadelta: mid-Holocene palaeoecology of the Orinoco Delta (NE Venezuela). Quat Sci Rev 221:105874

Muller J, Di Giacomo E, Van Erve AW (1987) A palynological zonation for the cretaceous, tertiary and quaternary of northern South America. Am Assoc Strat Palynol Contr Ser 19:7–76

Peros MC, Reinhardt EG, Schwarcz HP et al (2007) High-resolution paleosalinity reconstruction from Laguna de la Leche, north coastal Cuba, using Sr, O, and C isotopes. Palaeogeogr Palaeoclimatol Palaeoecol 245:535–550

Popescu S-M, Suc J-P, Fauquette S et al (2021) Mangrove distribution and diversity during three Cenozoic thermal maxima in the Notrhern Hemisphere (pollen records from the Arctic-North Atlantic-Mediterranean regions). J Biogeogr 48:2771–2784

Ramcharan EK, McAndrews JH (2006) Holocene development of coastal wetland at Maracas Bay, Trinidad, West Indies. J Coast Res 22:581–586

Rull V (2022) Responses of Caribbean mangroves to Quaternary climatic, eustatic and anthropogenic drivers of ecological change: a review. Plan Theory 11:3502

Rull V, Poumot C (1997) Eocene to Miocene palynocycles from western Venezuela and correlations with global eustatic cycles. Mem VII Cong Geol Venez II:343–349

Rull V, Vegas-Vilarrúbia T (1999) Surface palynology of a small coastal basin from Venezuela and potential paleoecological applications. Micropaleontology 45:365–393

Takayama K, Tateishi Y, Kajita T (2021) Global phylogeography of a pantropical mangrove genus *Rhizophora*. Sci Rep 11:7228

Triest L (2008) Molecular ecology and biogeography of mangrove trees towards conceptual insights on gene flow and barriers: a review. Aquat Bot 89:138–154

Van der Hammen TA (1963) palynological study on the Quaternary of British Guiana. Leidse Geol Mededel 29:125–168

Van der Hammen T, Wijmstra TA (1964) A palynological study on the tertiary and upper cretaceous of British Guiana. Leidse Geol Mededel 30:183–241

Westerhold T, Marwan N, Drury AJ et al (2020) An astronomically dated record of earth's climate and its predictability over the last 66 million years. Science 369:1383–1387

Wijmstra TA (1968) The identity of *Psilatricolporites* and *Pelliciera*. Acta Bot Neerl 17:114–116

Xu S, He Z, Zhang Z et al (2017) The origin, diversification and adaptation of a major mangrove clade (Rhizophoraceae) revealed by whole-genome sequencing. Natl Sci Rev 4:721–734

Yarincik KM, Murray RW (2000) Climatically sensitive aeolian and hemipelagic deposition in the Cariaco Basin, Venezuela, over the past 578,000 years: results from Al/Ti and K/Al. Paleoceanography 15:210–228

The *Pelliciera* Taxon Cycle: A Time-Continuum Integrative Synthesis

6

Abstract

The taxon cycle concept introduces the idea of a species undergoing repeated expansions and contractions in its range over time to sustain its core distribution indefinitely; otherwise, it faces extinction. Taxon cycles are commonly associated with tropical island faunas, examples from continental regions are scarce, and there is a lack of similar case studies for plants. Most taxon cycles have been recognized through phylogeographical studies, with limited empirical evidence from fossils. The fossil record of an emblematic Caribbean mangrove element, *Pelliciera*, provides empirical fossil evidence supporting the recurring expansion and contraction trends between its original core area and the whole Neotropics. Indeed, the observed recurrent patterns are consistent with the taxon cycle concept from biogeographical, chronological, and ecological perspectives. In a taxon cycle ontext, the *Pelliciera–Rhizophora* tradeoff during the EOT may be viewed as more than a facilitation process, as *Pelliciera* not only survived but have also had new unexpected possibilities for range expansion and speciation. The recurrent patterns of expansion and contraction observed in *Pelliciera* hold significant promise as the first taxon cycles that have been empirically recorded, and they appear to be the sole documented instance of taxon cycles in plants thus far. Whether the *Pelliciera* taxon cycle could represent a trend toward extinction may be relevant for its conservation.

Once the vast available Eocene-to-present fossil/living record of *Pelliciera* has come to light, it seems opportune to discuss the range shifts of this taxon in the framework of the long debated biogeographical and evolutionary concept of the taxon cycle. This approach would be important for two main reasons. First, *Pelliciera* provides unique fossil evidence for a taxon cycle and could constitute the first case

V. Rull, *Origin and Evolution of Caribbean Mangroves*, Ecological Studies 252,
https://doi.org/10.1007/978-3-031-57612-6_6

of this concept empirically documented to date for plants (Rull 2023). Second, the *Pelliciera* taxon cycle and the associated *Pelliciera–Rhizophora* ecological interactions provide a synthesis of the evolution of Caribbean mangroves under a perspective that perfectly illustrates the time-continuum integrative approach used in this book.

6.1 The Taxon Cycle

Wilson (1961) introduced the concept of the taxon cycle to characterize the ecological and evolutionary dynamics of species undergoing successive range expansions and contractions around its core distribution area, the headquarters, thus avoiding extinction. The taxon cycle comprises four main stages, as outlined by Ricklefs and Cox (1972) (Fig. 6.1). In the initial stage, high-density expanding populations, the "supertramps" (Diamond 1974), swiftly colonize new environments with minimal evolutionary differentiation across their range. Stage II sees a slowdown in expansion, accompanied by a significant increase in population differentiation, especially at range margins, with corresponding taxa labeled as "great speciators" (Diamond et al. 1976). Stage III involves biogeographical stasis, fragmentation with local extinction and incipient speciation, either initiating a new cycle or alternatively (Stage IV), a gradual decline in range size leading to progressive relictualization and eventual extinction (Pepke et al. 2019).

Originally, the taxon cycle concept was formulated for biogeographical patterns observed in island biotas (Wilson 1959). Although there have been suggestions that these cycles could also occur in continental environments (Ricklefs and Bermingham 2002), the primary focus remains on archipelagos. Various animal groups, such as crustaceans, insects, reptiles, and birds, particularly those found in tropical islands and archipelagos like the Caribbean Antilles, Indonesia, Melanesia, New Guinea, the Philippines, and Madagascar, have been studied under the taxon cycle model (MacLean and Holt 1979; Losos 1992; Jones et al. 2001; Cook et al. 2008; Simberlof and Collins 2010; Economo and Sarnat 2012; Jønson et al. 2014; Economo et al. 2015; Fuchs et al. 2016; Dalsgaard et al. 2018; Matos-Maraví et al. 2017; Oliver et al. 2018; O'Connell et al. 2019; Cozzarolo et al. 2019; Liu et al. 2020; Cognato et al. 2021). However, there is notably limited exploration of taxon cycles in continental areas. For example, a meta-analysis focusing on a widely distributed bird group (Campefagidae) spanning tropical regions of Asia, Africa, and Australia, covering both insular and continental regions, has been suggested to align with the taxon cycle concept (Pepke et al. 2019). Another example involves the diversification of Central American salamanders within the supergenus *Bolitoglossa* (Plethodontidae) (Garcia-Paris et al. 2000). Yet, comparable case studies for plants are lacking. Despite efforts to find empirical support for the predictions of the taxon cycle model in plants, no compelling examples have been discovered (Sheh et al. 2018). Only one molecular phylogenetic study on the plant tribe *Gaultherieae* (Ericaceae) was identified, suggesting that the evolution of fruit color in this tribe might undergo a process similar to a taxon cycle (Lu et al. 2019).

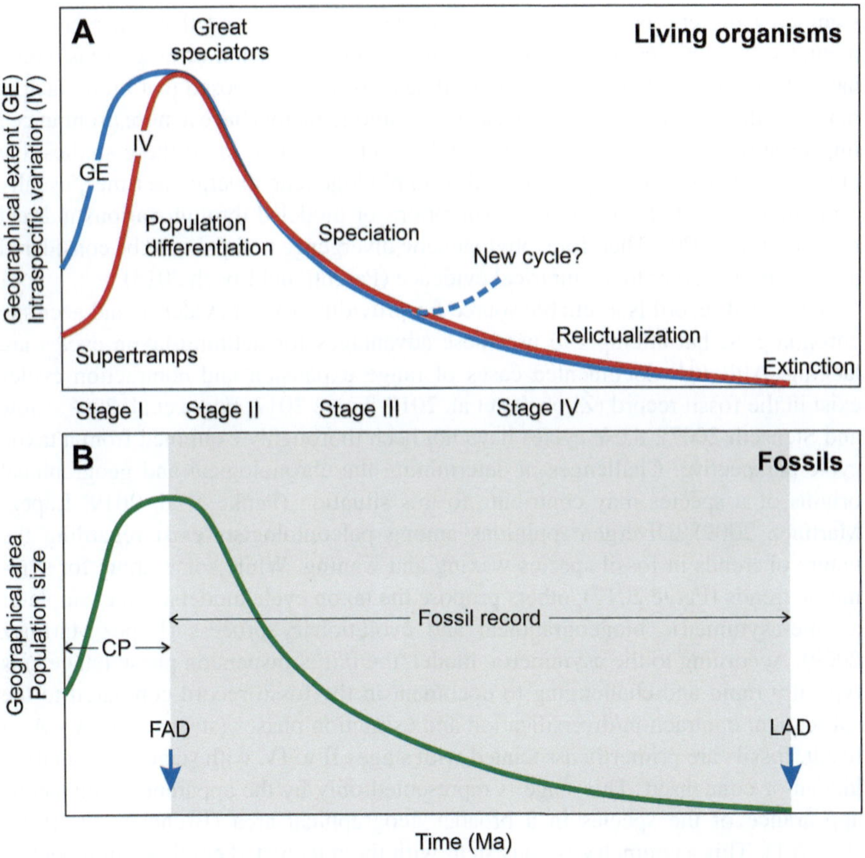

Fig. 6.1 (**a**) The four stages of the taxon cycle, as defined for living species (redrawn from Pepke et al. 2019). (**b**) The fossil expression of the taxon cycle (gray area) considering the asymmetric model (López-Martínez 2009). CP, condensed phase (usually not observed in the fossil record); FAD, first appearance datum; LAD, last appearance datum

The occurrence of taxon cycles was attributed by its proponents to changes in biotic interactions, specifically competition and predation, rather than shifts in environmental factors (Wilson 1961; Ricklefs and Cox 1972). Subsequently, it was recognized that the estimated duration of taxon cycles in Caribbean birds, determined through molecular phylogenetic analyses, extended over approximately 10^5–10^7 years (Ricklefs and Bermingham 2002). This duration significantly surpasses that of most cyclic climatic drivers, especially the Pleistocene glacial–interglacial cycles, which have periods of 0.02–0.1 million years (Hays et al. 1976). As a result, the idea of a biotic origin and control gained further support. Other investigations, utilizing similar methods, calculated taxon cycle periodicities of around five million years for Indo-Pacific birds (Pepke et al. 2019; Jønsson et al. 2014), strengthening the notion that Pleistocene climatic cycles might not have been significant drivers. Nevertheless, these researchers suggested that other environmental shifts with longer periodicities, such as plate collision or orogenesis, could have played a role (Pepke et al. 2019). It has also been proposed that biotic factors mainly influence the expansion phase, while abiotic factors have a more pronounced impact during the retraction phase (Žliobaitė et al. 2017). In all these studies, the duration of taxon cycles was deduced from phylogenetic divergence times, usually estimated using molecular clock assumptions or modeled through various indirect methods (Ho 2020). Therefore, phylogenetic divergence times should be considered as hypotheses rather than empirical evidence (Parenti and Ebach 2013).

The fossil record is a reliable source for providing robust evidence and accurate chronologies, but attempts to use these advantages for defining taxon cycles are lacking. Although documented cases of range expansion and contraction cycles exist in the fossil record (Žliobaitė et al. 2017; Foote 2017; Foote et al. 2007; Liow and Stenseth 2007), these cycles have not been thoroughly examined from a taxon cycle perspective. Challenges in determining the chronological and geographical origins of a species may contribute to this situation (Pepke et al. 2019; López-Martínez 2009). Divergent opinions among paleontologists exist regarding the nature of trends in fossil species waxing and waning. While some argue for symmetric trends (Foote 2017), others propose the taxon cycle model as an example of a time-asymmetric biogeographical and evolutionary process (López-Martínez 2009). According to the asymmetric model, the initial dispersion phase (stage I) is typically rapid and challenging to document in the fossil record compared to the subsequent contraction/diversification and extinction phases (stages II to IV). As a result, fossils are primarily associated with stages II to IV, with stage I being either hidden or condensed. This stage is represented only by the apparent simultaneous appearance of the species in a broader geographical area (Brunet et al. 1995) (Fig. 6.1). This asymmetry is consistent with the punctuated equilibrium model of evolution, considering it an intrinsic characteristic of the fossil record and the evolutionary process itself rather than a perceived flaw (Eldredge and Gould 1972; Gould and Eldredge 1977). According to the asymmetric model, the last appearance datum (LAD) reliably indicates a fossil extinction, while the first appearance datum (FAD) does not accurately reflect its actual time and place of origin but rather its initial spreading (Fig. 6.1).

6.2 The Case of *Pelliciera*

Pelliciera has conventionally been considered a monotypic genus, but recent findings by Duke (2020) have resulted in its reclassification into two distinct species: *P. rhizophorae* and *P. benthamii*. These species are limited to a specific region along the Caribbean and Pacific coasts of Central America and northwestern South America, referred to in this context as the present *Pelliciera* range (PPR) (Fig. 6.2). Despite its current distribution, *Pelliciera* was more widespread across the Neotropics after losing dominance in the Eocene (Chap. 4) than it is today, posing a significant biogeographical puzzle (Wijmstra 1968; Graham 1977, 1995; Rull 1998,

2001). Various explanations have been suggested to understand the reduction in *Pelliciera* from the Miocene to the present, including factors such as climatic and/or salinity stress, sea level changes, or competition with *Rhizophora* (Wijmstra 1968; Fuchs 1970; Graham 1977, 1995; Jiménez 1984; Rull 1998, 2001). However, it is important to consider some methodological shortcomings. First, the notion of a post-Miocene range reduction was based on the analysis of a few fossil records, and this perspective has endured without reassessment using an updated fossil database. Second, the narrative of Cenozoic *Pelliciera* range shifts is incomplete, as evidence from the pre-Miocene period has not been scrutinized with the same criteria to achieve a comprehensive spatiotemporal fossil record. Third, most prior hypotheses relied on pollen records that were qualitative and semiquantitative, and as demonstrated in Chap. 3, quantitative data are crucial for accurately documenting and understanding the evolution of Neotropical mangroves (Rull 2022a).

The utilization of the CARMA dataset has brought to light a more complete view of the biogeographical shifts of *Pelliciera* through time and revealed a prolonged expansion-contraction pattern compatible with the taxon cycle concept proposed by Wilson (1961). The following discussion delves into ecological and evolutionary aspects linked to this long-term cycle, in light of the predictions of the taxon cycle model. Additionally, environmental drivers outlined in Chap. 4 (Sect. 4.3.2) that

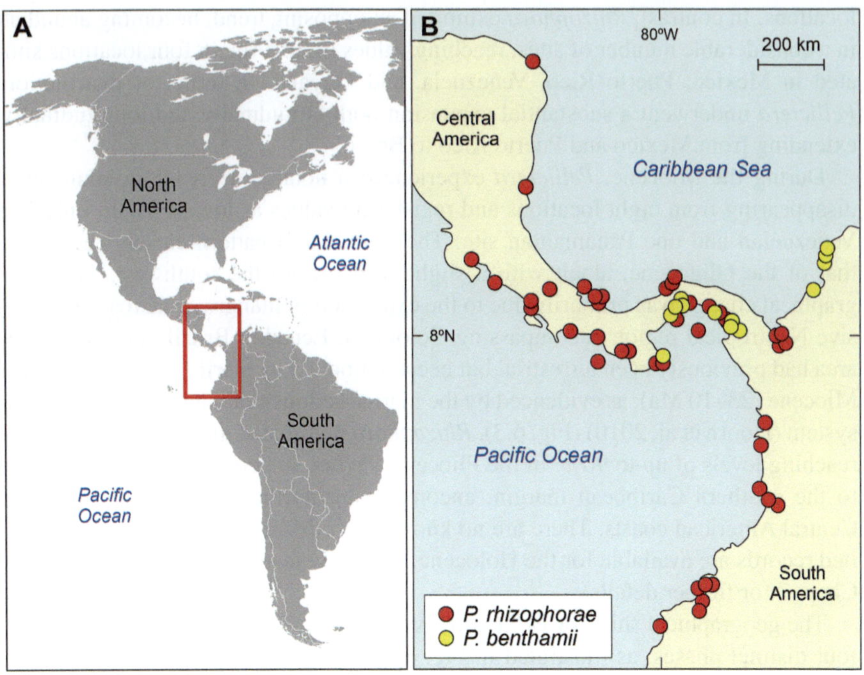

Fig. 6.2 Present distribution of *Pelliciera*. (**a**) Map of the Americas showing the distribution area of *Pelliciera* (red box). (**b**) Close-up showing the distribution of the two *Pelliciera* species (Modified from Duke 2020)

may play a role in the *Pelliciera* cycle are explored. The current state of *Pelliciera* populations, compared to their past biogeographical history, is assessed to anticipate potential future developments. The account also includes *Rhizophora*, as a fundamental element to understand major ecological and evolutionary trends involved in the *Pelliciera* cycle. The raw information used in this discussion has been taken from previous chapters, and for this reason, some repetitions are inevitable for contextual refreshment. To offer a comprehensive Neotropical perspective, the CARMA dataset has been enhanced with the data available from other localities beyond the Caribbean region (Table 6.1).

6.3 Biogeographical Shifts

The findings from the extended CARMA dataset (Tables 4.2 and 6.1) are summarized in Fig. 6.3. In the Eocene, *Pelliciera* was prevalent or abundant in most locations where it was identified, with maximum percentages reaching up to 60% (Table 4.2). Its distribution was limited to NW South America (currently Colombia and Venezuela), along with one site in Panama and another in Jamaica. *Rhizophora*, on the other hand, was found in only six locations, with a noticeable presence (≤10%) only in Panama. Moving into the Oligo-Miocene, there was a significant decrease in the abundance of *Pelliciera*, dropping to values below 5%, except in two Venezuelan locations. In contrast, *Rhizophora* exhibited an opposing trend, becoming abundant in a considerable number of sites, reaching values of 50–90% in four locations situated in Mexico, Puerto Rico, Venezuela, and Guyana. In terms of distribution, *Pelliciera* underwent a substantial expansion both latitudinally and longitudinally, extending from Mexico and Puerto Rico to Brazil.

During the Miocene, *Pelliciera* experienced a notable decrease in abundance, disappearing from eight locations and registering values as low as 3% in only five Venezuelan and one Panamanian site. The distribution pattern closely resembled that of the Oligocene, albeit with a slight shift toward the southwest. This geographical change was primarily due to the expansion of mangroves across an extensive Neotropical region encompassing Colombia, Ecuador, Brazil, and Peru. This area had previously been terrestrial but became flooded by marine waters during the Miocene (23–10 Ma), as evidenced by the marine sedimentary deposits of the Pebas system (Hoorn et al. 2010) (Fig. 6.3). *Rhizophora* dominated most Miocene records, reaching levels of up to 90%. In the Pliocene, *Pelliciera* persisted but was confined to the southern Caribbean margin, encompassing northern South American and Central American coasts. There are no known records for the Pleistocene, and limited records are available for the Holocene, primarily restricted to the PPR (refer to Chap. 7 for further details).

The geographical shifts of *Pelliciera*, as depicted in Fig. 6.3, can be divided into four distinct phases, as illustrated in Fig. 6.4. During the initial phase (Eocene to Oligocene), the range expanded across most of the Neotropics, but populations experienced a significant reduction, leading to a more widespread yet "diluted" distribution. This phase is referred to as "thinning expansion" here, as the term

Table 6.1 Neotropical localities outside the Caribbean region, and therefore absent in the CARMA dataset, utilized in this study, indicating the presence (+), absence (–) and abundance (%) data for *Pelliciera* and *Rhizophora*, according to the original references. E, Early; M, Middle

Site	Map	Country	Age	*Pelliciera*	*Rhizophora*	References
Alliance	Ac	Surinam	Miocene	+	+	Wijmstra (1969)
Mariñame	Mñ	Colombia	E/M Miocene	+	≤70%	Hoorn (1994)
Chorros	Cr	Colombia	Miocene	+	≤20%	Hoorn et al. (2022)
La Frontera	Fn	Peru	Miocene	+	–	Parra et al. (2020)
105-AM	AM	Colombia	E/M Miocene	+	+	Jaramillo et al. (2017)
Pará	Pr	Brazil	Miocene	≤2%	≤40%	Antonioli et al. (2015); Aguilera et al. (2017)
Amazon	Az	Brazil	Oligo-Miocene	+	+	De Boer et al. (1965)
Maranhão	Mh	Brazil	Oligo-Miocene	+	+	Regali et al. (1974)
Barreirinhas	Bh	Brazil	Oligo-Miocene	+	+	Regali et al. (1985)
Coastal Plain	CP	Surinam	Oligo-Miocene	+	+	Amstelveen (1971)

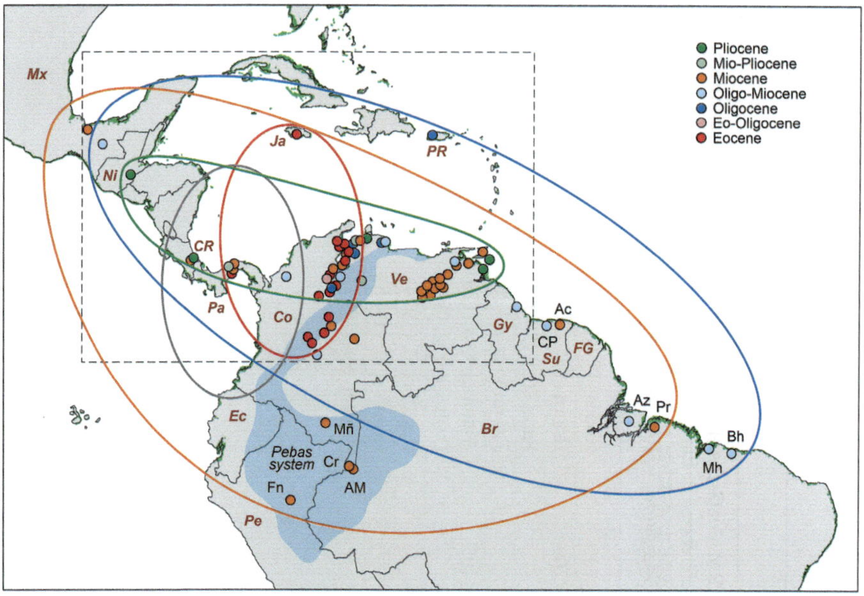

Fig. 6.3 Neotropical sites with *Pelliciera* pollen sorted chronologically. Localities beyond the Caribbean region, as defined in this study (dashed box), according to Table 6.1. The approximate extension of the Miocene marine incursion represented by the Pebas system (in blue) has been taken from Hoorn et al. (2010). The distribution area for extant *Pelliciera* is taken from Fig. 6.2 and encircled by a gray line. The present countries with fossil *Pelliciera* sites are indicated: Br, Brazil; Co, Colombia; CR, Costa Rica; Ec, Ecuador; FG, French Guiana; Gy, Guyana; Ja, Jamaica; Mx, Mexico; Ni, Nicaragua; Pe, Peru; PR, Puerto Rico; Su, Surinam; Ve, Venezuela

"dilution" has been used in biogeography for other processes (Keesing et al. 2006, 2010). The second phase (Oligocene to Miocene) involved a displacement, where the range slightly shifted to the northwest, with no notable differences compared to the Oligocene. The third phase (Miocene to Pliocene) saw a substantial contraction of the range toward the southern Caribbean margin. Finally, in the fourth phase (Pliocene to the present), the range did not undergo further reductions but exhibited a longitudinal contraction around Central America and a southward migration toward the equator, constituting a spatial reorganization. In summary, the *Pelliciera* range significantly expanded from its original headquarters in northwestern South America until reaching a Miocene maximum, covering most of the Neotropics, before undergoing a retraction that culminated in the present distribution, similar in location and extent to the Eocene range. This biogeographical loop was accompanied by a notable decline in *Pelliciera* populations, which was abrupt during the EOT and gradual from the Oligocene to the present. Considering that the Eocene *Pelliciera* mangroves reached their maximum extent in the Lutetian (Middle Eocene) (Rull 2022a), the minimum duration of the entire *Pelliciera* biogeographical loop is estimated to be approximately 45 million years.

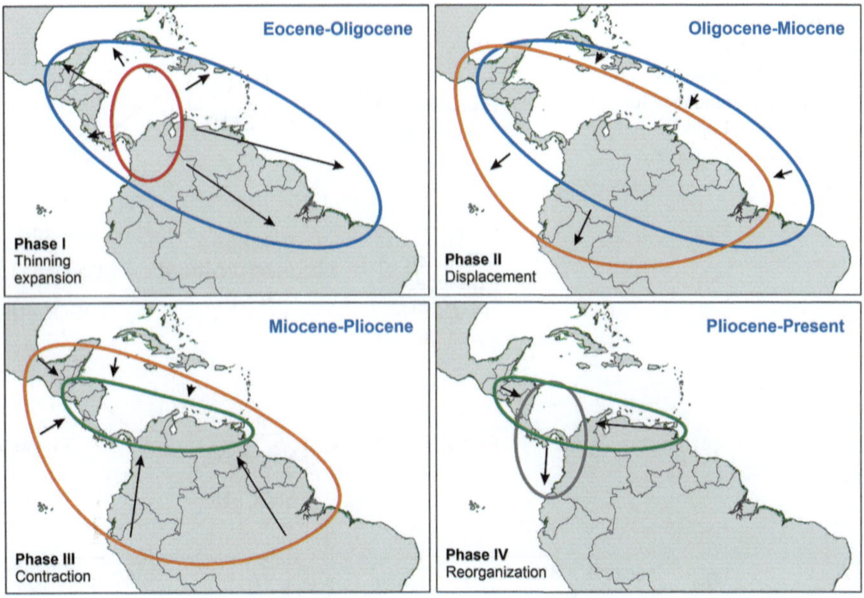

Fig. 6.4 *Pelliciera* range shifts from the Eocene to the present split into four chronological phases. (**a**) Phase I, Eocene to Oligocene thinning expansion; (**b**) Phase II, Oligocene to Miocene displacement; (**c**) Phase III, Miocene to Pliocene contraction; (**d**) Phase IV, Pliocene to present reorganization

6.4 The *Pelliciera* Loop

The *Pelliciera* loop (Fig. 6.5), as revealed in the above fossil records, provides clear empirical support for a cycle of expansion and contraction consistent with the notion of a taxon cycle, as proposed by Wilson (1961). This section delves deeper into whether this cycle genuinely represents a taxon cycle, examining it in greater detail and giving special attention to potential ecological and evolutionary implications. Previously, it is important to highlight some peculiar features of the *Pelliciera* cycle.

First, the Eocene to Miocene expansion phase lasted much longer than anticipated by the asymmetric model, as proposed by López-Martínez (2009). This model suggested that Phase I was too rapid to be discernible in the fossil record (Fig. 6.1), and it was expected to be condensed within the Eocene when *Pelliciera* had already present across NW South America, Central America, and the Greater Antilles (Fig. 6.3). However, when the Eocene records are further split into Early, Middle, and Late Eocene (Table 4.2), a minor intra-Eocene contraction-expansion cycle is revealed (Fig. 6.6). According to this cycle, *Pelliciera* likely originated in the Early Eocene in western Venezuela, experienced a rapid expansion peaking in the Middle Eocene, and then retreated in the Late Eocene to an area similar but not identical to its original range (Fig. 6.6). Consequently, the actual Stage I would have occurred between the Early and Middle Eocene. While determining the exact duration of this

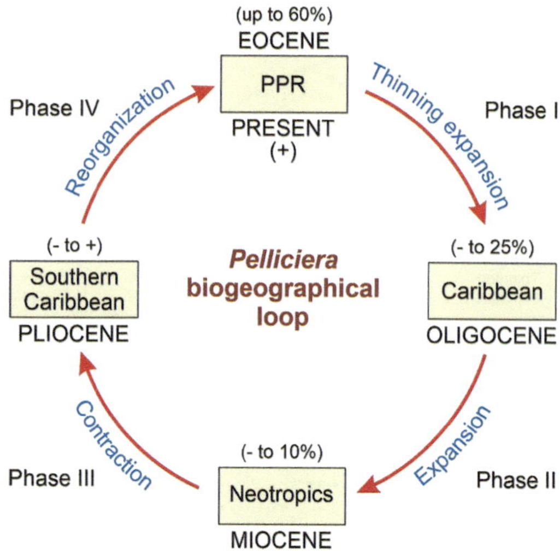

Fig. 6.5 The *Pelliciera* biographical loop indicating the ranges of fossil pollen abundance for each geological epoch. PPR, present *Pelliciera* range

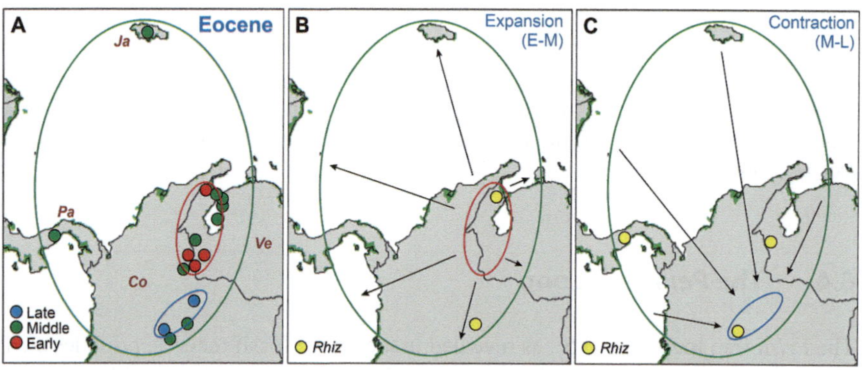

Fig. 6.6 The intra-Eocene *Pelliciera* loop. (**a**) Eocene localities, as depicted in Fig. 6.3, subdivided into Early, Middle, and Late Eocene. (**b**) Early–Middle Eocene expansion. (**c**) Middle–Late Eocene contraction. Countries: Co, Colombia; Ja, Jamaica; Pa, Panama; Ve, Venezuela. *Rhiz, Rhizophora* occurrences

cycle proved challenging, a maximum estimate could span the entire Eocene epoch, covering approximately 2×10^7 years. Applying the same logic, the intra-Eocene Stage I would have endured a maximum of 10^7 years.

Second, the possibility of population differentiation and potential speciation, as predicted by the taxon cycle model, can be tentetively explored through an examination of pollen morphology. Previous studies (Castillo-Cárdenas et al. 2014, 2015; Duke 2020) have demonstrated the utility of pollen morphology in distinguishing current *Pelliciera* species and subspecific taxa. Key distinguishing factors include size and the sculpturing of the outer pollen wall (exine). The fossil *Pelliciera* pollen, spanning the Eocene to the Pliocene, also displays considerable morphological variation in these defining characteristics (Germeraad et al. 1968; Frederiksen 1985; Lorente 1986; Muller et al. 1987; Jaramillo and Dilcher 2001). Unfortunately, there are no systematic studies that allow for differentiation among potential specific or intraspecific fossil taxonomic categories. While future studies should address this issue in more depth, the existing fossil evidence does not rule out the possibility of distinct *Pelliciera* species or subspecies.

Third, the application of the niche conservatism concept suggests that ancient *Pelliciera* species likely exhibited a narrow ecological range, mirroring the characteristics of their present-day counterparts. This is particularly evident in the Eocene, where the *Pelliciera* range did not extend beyond tropical warm and humid climates, akin to its current distribution (Chap. 4, Sect. 4.3.2). Notably, the Eocene epoch experienced significantly higher temperatures than the present, with global averages ranging from approximately 8 to 14 °C above current temperatures (Fig. 1.13). This implies that the fossilized *Pelliciera* species thrived in macrothermal conditions during that time.

Fourth, the earliest documented occurrences of *Rhizophora* date back to the Middle Eocene, continuing through the Late Eocene in sporadic appearances. A noteworthy exception is a site in Panama, where *Rhizophora* constituted 10% of the pollen assemblage in the Late Eocene (Fig. 6.6). These findings mark the initial phases of *Rhizophora* colonization of the Neotropics, coinciding with recent molecular phylogeographical studies that trace the origin of this genus in the IWP region, with pantropical distribution by the Mid–Late Eocene (Takayama et al. 2021). According to the same studies, the possibility exists that *Rhizophora* migrated to the Neotropics through the Atlantic Ocean via the Tethys seaway (Lo et al. 2014). However, the support for or rejection of this hypothesis cannot be determined through paleogeographical reconstructions and fossil pollen records alone. This uncertainty arises because both the Atlantic and Pacific seaways to the Caribbean were open for dispersal throughout the entire Eocene (Mann 2021; Romito and Mann 2020), and the fossil records do not exhibit a clear Atlantic or Pacific pattern. The only exception is the Panamanian site, which was located in the former Pacific island arc of the western Caribbean plate margin during the Eocene. This exceptional site introduces the possibility of a Pacific dispersal pathway, but additional research is necessary for a comprehensive evaluation.

Coming back to the fossil *Pelliciera* loop, the expansion from the Oligocene to the Miocene (thinning expansion) originated in a small region in northwestern South America where the Eocene and current ranges intersected (Figs. 6.4 and 6.6). From its initiation, this expansion closely followed the spread of *Rhizophora*, the predominant mangrove tree, and the shift in dominance appears to have been influenced significantly by the environmental disruptions during the EOT, coupled with biotic interactions. The survival and geographical expansion of *Pelliciera*, despite facing intense competition, can be explained by a combination of facilitation and niche segregation (see Chap. 4 for more details). It is worth adding that eurytopic generalists such as *Rhizophora* are able at invading the habitats of stenotopic specialists such as *Pelliciera*. However, specialists tend to be more efficient within a limited range of conditions (microhabitat) in which they can thrive (Futuyma and Moreno 1988). In this context, the shift from *Pelliciera* to *Rhizophora* dominance, without outright competitive exclusion, can be interpreted as a trade-off. The newly arrived generalist (*Rhizophora*) outcompetes the resident specialist (*Pelliciera*) in terms of dominance. In return, it facilitates a microhabitat for the ecological loser (sensu Fernández-Palacios et al. 2021), which "accepted" to play a secondary role in the ecosystem for survival.

In this way, *Pelliciera* not only secured its survival but also gained the chance to extend its presence across the entire Neotropical region, surpassing the boundaries of its Eocene macroenvironmental niche during the Oligocene and the Miocene epochs. This expansion would likely have been improbable without the assistance of *Rhizophora*. This is suggested by *Pelliciera* being consistently confined to its preferred PPR zone, referred to as headquarters by Wilson (1961), throughout the entire Eocene when *Rhizophora* was largely absent. Given the limited dispersal ability of *Pelliciera*, mainly occurring through short-range movements via coastal currents, its expansion likely occurred through diffusion—a gradual migration across favorable terrains—rather than long-distance dispersal, which involves crossing inhospitable landscapes (Pielou 1977). This suggests that *Rhizophora* played a vital role as the primary colonizer. Once the *Rhizophora* mangrove community was established, creating a suitable microhabitat for *Pelliciera*, this taxon could then establish itself. Consequently, during its peak expansion in the Oligo–Miocene period, *Pelliciera* probably existed as a dispersed network of small populations confined to favorable microhabitats, as indicated by fossil pollen records. Indeed, while these records were widespread, they displayed low pollen percentages when present. This spatial arrangement and ecological dynamics are consistent with the concept of microrefugia, particularly the diffuse type, promoting genetic differentiation among populations of the same species (Rull 2009). Despite initially being an ecological loser, *Pelliciera* might have encountered unprecedented opportunities for diversification, ultimately becoming an evolutionary winner, as described by Fernández-Palacios et al. (2021).

From an environmental perspective, *Pelliciera* experienced a period of extended climatic stability, characterized by minor fluctuations, during its expansion (Westerhold et al. 2020) (Fig. 1.11). This indicates that the growth phase was heavily influenced by biotic interactions. It is crucial to note that the Oligo–Miocene spread, whose duration is challenging to pinpoint due to limited temporal resolution in the fossil record, marked the second expansion of *Pelliciera*. As quoted above, the initial expansion occurred during the Early–Middle Eocene within its original range, followed by a contraction in the Late Eocene. Consequently, *Pelliciera* has undergone at least two cycles of expansion and contraction since the Eocene. The reasons behind the occurrence of the first cycle are unknown, but the second cycle was likely triggered by the arrival of *Rhizophora* in the Late Eocene. *Rhizophora* not only provided new opportunities for spreading and diversification but also played a key role in facilitating the process as a niche builder and an indirect dispersal agent, essentially acting as an ecological "nurse" for *Pelliciera*. This hypothesis should be tested with further studies, especially in relation with the involved ecophysiological mechanisms. The possibility of *Pelliciera* facing extinction during its range contraction, as predicted by the taxon cycle model, is uncertain. If extinction was imminent, *Rhizophora* might have created conditions for *Pelliciera* to overcome this bottleneck. Therefore, in ecological and evolutionary dynamics, *Rhizophora*, initially a competitor, would not only act as a "rescuer" for *Pelliciera* but also serve as its evolutionary "sponsor/benefactor."

The Miocene–Pliocene contraction followed a significant cooling event known as the Middle Miocene Cooling Transition (MMCT) (Westerhold et al. 2020) (Fig. 1.11). The fossil pollen record showed no noticeable changes, indicating that climatic factors may have played a more prominent role than biotic factors in causing this contraction. The confinement of *Pelliciera* to the southern Caribbean margin during the Pliocene can be explained by the local extinction of all Miocene *Pelliciera* populations beyond this region, which agrees with predictions from the taxon cycle theory. It seems plausible that, despite the protection of *Rhizophora*, *Pelliciera* could not withstand a second cooling and survived only in the part of its range where favorable temperature and precipitation persisted. This idea is supported by the evident latitudinal component observed in the contraction. The rearrangement between the Pliocene and the present poses two main puzzles. First, the current distribution is based on direct records of existing populations, while past distributions relied on inferences from fossil pollen assemblages susceptible to various taphonomic processes that do not affect living plants. Second, a gap between the Pliocene and the present, as highlighted by Rull (2022b) (see also Chap. 7), hinders our understanding of how Pleistocene glaciations, the coldest phases of the entire Cenozoic era, affected the range of *Pelliciera*. Considering the macrothermic nature of this taxon, it is expected that glaciations had a significant impact on its populations, although currently, there is no evidence supporting this possibility.

One notable characteristic of the past millennium, absent in earlier geological times, is the presence of humans. The impact of humans on *Pelliciera* remains largely unclear, with insights mainly emerging in recent decades. *P. rhizophorae*, is currently classified as "Vulnerable" on the IUCN (International Union for the Conservation of Nature) Red List due to its limited (500–2000 km^2) and fragmented distribution, facing a significant risk of extinction in its natural habitat (Polidoro et al. 2010; Blanco et al. 2012; Bhowmik et al. 2022). The expansion of urban areas poses a substantial threat to *Pelliciera*, resulting in the loss and fragmentation of its habitat (Blanco-Libreros and Ramírez-Ruiz 2021). The available evidence suggest that human activities have not necessarily reduced the overall distribution of *Pelliciera* but have induced severe fragmentation, impacting population sustainability and heightening vulnerability to extreme events and global warming (Blanco-Libreros and Ramírez-Ruiz 2021). Concerns arise in light of the taxon cycle, questioning whether *Pelliciera* is nearing complete extinction and examining how human actions may hasten this process. If the IUCN adopts this perspective, there is a possibility of reclassifying *Pelliciera* as "Critically Endangered" and designating it as a conservation priority. The likelihood of another evolutionary rescuer similar to *Rhizophora* emerging to prevent *Pelliciera* extinction and initiate a third cycle appears improbable. Although humans could potentially contribute to conservation and restoration efforts, ethical dilemmas surface regarding the artificial preservation of a taxon likely facing natural extinction (Rull 2023).

6.5 Summary

The taxon cycle delineates the biogeographical and evolutionary dynamics of species, allowing them to evade extinction through recurring expansions and contractions in their core distribution area, commonly referred to as the headquarters. Although initially formulated for island faunas, this concept has been extended to continental environments. It encompasses four primary stages: (I) expansion marked by colonization of new environments without discernible evolutionary differentiation; (II) diminishing expansion rates coupled with increasing population differentiation; (III) geographical stasis and population fragmentation due to local extinctions fostering speciation; and (IV) range contraction, relictualization, and the initiation of a new cycle or extinction. Utilizing phylogenetic divergence times, the duration of a taxon cycle is estimated to range from 10^5 to 10^7 years, with biotic drivers influencing the expansion phase and abiotic climatic/tectonic drivers impacting the contraction phase. Most identified taxon cycles rely on extant biogeographical patterns and molecular phylogeographical evidence, and there remains a scarcity of unequivocal fossil evidence, particularly for plants.

Based on the examination of fossil pollen discussed in the previous chapters, *Pelliciera* underwent a prolonged expansion and contraction cycle, coinciding with the taxon cycle concept, from the Eocene to the present. The central hub for *Pelliciera* was located in NW South America, marking the initial appearance of Caribbean mangroves. The major expansion phase occurred from the Oligocene to the Miocene, reaching its widest distribution across the Neotropics. Subsequently, the contraction phase began, leading to the confinement of *Pelliciera* to the coasts of Central and northern South America between the Pliocene and the present. The current geographical range resembles that of the Eocene. This entire cycle spanned approximately 4×10^7 years and was further divided into two shorter taxon cycles, the first from the Early to the Late Eocene and the second from the Oligocene to the present, each lasting approximately 2×10^7 years.

The Eocene cycle was initiated and completed in the vicinity of the *Pelliciera* headquarters. The Oligo-Miocene *Pelliciera* expansion, marking the onset of the second cycle, coincided with the spread of *Rhizophora* across the Neotropics. This expansion, termed "thinning expansion," saw *Pelliciera* extend its range but be confined to small microhabitats within the *Rhizophora*-dominated mangrove landscape. Consequently, *Rhizophora*, emerging as the ecological winner, outcompeted *Pelliciera*, the ecological loser, becoming the predominant mangrove-forming tree in the Neotropics. Simultaneously, the presence of *Rhizophora* facilitated the expansion and persistence of *Pelliciera*, positioning it as the evolutionary winner that might have faced extinction during phase IV (Late Eocene) without this ecological dynamic. In this context, the ecological interactions between *Pelliciera* and *Rhizophora* can be viewed as a progressive transition from competition to facilitation to sponsorship. This progression has allowed the coexistence and survival of both species through niche segregation, evident in the shift in dominance and the restriction of *Pelliciera* to intra-mangrove microrefugia.

From an environmental perspective, the cooling associated with the EOT may have played a role in enhancing the competitive advantage of the widespread and vagile *Rhizophora* over the restricted and specialized *Pelliciera*. The latter, which thrives in warm and humid near-equatorial climates and has limited dispersal capabilities, experienced a notable reduction in its range during the significant cooling event known as the MMCT. During the Oligo-Miocene expansion, both *Rhizophora* and *Pelliciera* benefited from an extended period of climatic stability. However, the subsequent contraction in the Mio-Pliocene occurred after a pronounced cooling phase (MMCT). This suggests that biotic interactions might have been more influential during the expansion, while external environmental factors played a more crucial role in the retraction phase, as predicted by the taxon cycle model. The substantial reduction in the range of *Pelliciera* during the Pliocene can be attributed to the increasing Miocene extinction, characteristic of phase III. This was likely a consequence of the combined effects of the MMCT cooling event and the overall Pliocene cooling trend. The impact of Pleistocene glacial–interglacial cycles remains uncertain due to a lack of records. Additionally, the scarcity of Holocene records hinders our understanding of the influence of human activities on *Pelliciera* biogeography. However, it is known that *Pelliciera* populations have declined and become fragmented in recent decades, leading to its classification as "Vulnerable" on the IUCN Red List of Threatened Species. In the context of the taxon cycle, questions arise about the potential imminent extinction of *Pelliciera* (end of phase IV) and how human activities might expedite this process. It is challenging to envision the emergence of another evolutionary rescuer akin to *Rhizophora*, which prevented *Pelliciera* extinction and facilitated its expansion, initiating a potential third cycle.

References

Aguilera O, Oliveire G, Tadeu R et al (2017) Neogene proto-Caribbean porcupinefishes (Diodontidae). PLoS One 12:e0181670

Amstelveen ALE (1971) A palynostratigraphical correlation of seven test wells in the Coastal Plain of Surinam. Geol Mijnb 21:177–182

Antonioli L, de Araújo V, Dino R (2015) Palynology of carcinolites and limestones from the Baunilha Grande Ecofacies of the Pirabas Formation (Miocene of Pará state, northeastern Brazil). J S Am Earth Sci 62:134–147

Bhowmik AK, Padmanaban R, Cabral P et al (2022) Global mangrove deforestation and its interacting social-ecological drivers: a systematic review and synthesis. Sustain For 14:4433

Blanco JF, Estrada EA, Ortiz LF et al (2012) Ecosystem-wide impacts of deforestation in mangroves: the Urabá Gulf (Colombian Caribbean) case study. ISRN Ecol 2012:958709

Blanco-Libreros JF, Ramírez-Ruiz K (2021) Threatened mangroves in the Anthropocene: Habitat fragmentation in urban coastalscapes of *Pelliciera* spp. (Tetrameristaceae) in northern South America. Front Mar Sci 8:670354

Brunet M, Beauvilain A, Coppens Y et al (1995) The first Australopithecine 2500 km west of the Rift Valley (Chad). Nature 378:273–275

Castillo-Cárdenas MF, Díaz-Gonzales F, Cerón-Souza I et al (2014) Jumping a geographical barrier: diversification of the mangrove species *Pelliciera rhizophorae* (Tetrameristaceae) across the Central American Isthmus. Tree Genet Genomes 11:1–11

Castillo-Cárdenas MF, Ramirez-Silva JA, Sanjur O (2015) Evidence of incipient speciation in the Neotropical mangrove *Pelliciera rhizophorae* (Tetrameristaceae) as revealed by molecular, morphological, physiological and climatic characteristics. Bot J Linn Soc 179:499–510

Cognato AI, Smith SM, Jordal BH (2021) Patterns of host tree use within a lineage of saproxlic snout-less weevils (Coleoptera: Curculionidae: Scolytinae: Scolytini). Mol Phylogenet Evol 159:107107

Cook BD, Pringle CM, Hughes JM (2008) Molecular evidence for sequential colonization and taxon cycling in freshwater decapod shrimps on a Caribbean Island. Mol Ecol 17:1066–1075

Cozzarolo C-S, Balke M, Buerki S et al (2019) Biogeography and ecological diversification of a mayfly clade in New Guinea. Front Ecol Evol 7:233

Dalsgaard B, Kennedy J, Simmons B et al (2018) Trait evolution, resource specialization and vulnerability to plant extinctions among Antillean hummingbirds. Proc R Soc B 285:20172754

De Boer NP, Van der Hammen T, Wijmstra TA (1965) A palynological study on the age of some borehole samples from the Amazonas Delta area, northeastern Brazil. Geol Mijnb 44:254–258

Diamond JM (1974) Colonization of exploded volcanic islands by birds: the supertramp strategy. Science 184:803–806

Diamond JM, Gilpin ME, Mayr E (1976) Species-distance relation for birds of Solomon archipelago, and paradox of great speciators. Proc Natl Acad Sci USA 73:2160–2164

Duke NC (2020) A systematic revision of the vulnerable mangrove genus *Pelliciera* (Tetrameristaceae) in equatorial Ammerica. Blumea 65:107–120

Economo EP, Sarnat EM (2012) Revisiting the ants of Melanesia and the taxon cycle: historical and human-mediated invasions of a tropical archipelago. Am Nat 180:E1–E16

Economo EP, Sarnat EM, Janda M (2015) Breaking out biogeographical modules: range expansion and taxon cycles in the hyperdiverse ant genus Pheidole. J Biogeogr 42:2289–2301

Eldredge N, Gould SJ (1972) Punctuated equilibria: an alternative to phyletic gradualism. In: Schopf TJM (ed) Models in paleobiology. Freeman Cooper, San Francisco, pp 193–223

Fernández-Palacios JM, Otto R, Borregaard MK et al (2021) Evolutionary winners are ecological losers among oceanic island plants. J Biogeogr 48:2186–2198

Foote M (2017) Symmetric waxing and waning of marine invertebrate genera. Paleobiology 33:517–529

Foote M, Crampton JS, Beu AG et al (2007) Rise and fall of species occupancy in Cenozoic fossil mollusks. Science 318:1131–1134

Frederiksen NO (1985) Review of early Tertiary sporomorph paleoecology. Am Assoc Strat Palynol Contr Ser 19:1–92

Fuchs HP (1970) Ecological and palynological notes on *Pelliciera rhizophorae*. Acta Bot Neerl 19:884–894

Fuchs J, Lemoine D, Parra JL et al (2016) Long-distance dispersal and inter-island colonization across the western Malagasy Region explain diversification in brush-warblers (Passeriformes: Nesillas). Biol J Linn Soc 119:873–889

Futuyma DJ, Moreno G (1988) The evolution of ecological specialization. Annu Rev Ecol Syst 19:207–233

Garcia-Paris M, Good DA, Parra-Olea G et al (2000) Biodiversity of Costa Rican salamanders: implications of high levels of genetic differentiation and phylogeographic structure for species formation. Proc Natl Acad Sci USA 97:1640–1647

Germeraad JH, Hopping CA, Muller J (1968) Palynology of Tertiary sediments from tropical areas. Rev Palaeobot Palynol 6:189–348

Gould SJ, Eldredge N (1977) Punctuated equilibria: the tempo and mode of evolution reconsidered. Paleobiology 3:115–151

Graham A (1977) New records of *Pelliciera* (Theaceae/Pelliceraceae) in the Tertiary of the Caribbean. Biotropica 9:48–52

Graham A (1995) Diversification of Gulf/Caribbean mangrove communities through Cenozoic time. Biotropica 27:20–27

Hays JD, Imbrie J, Shackleton NJ (1976) Variations in Earth's orbit: pacemaker of ice ages. Science 194:1121–1132

Ho SYW (2020) The molecular evolutionary clock. Theory and practice. Springer Nature, Cham

Hoorn C (1994) Fluvial palaeoenvironments in the intracratonic Amazonas Basin (Early Miocene-early Middle Miocene, Colombia). Palaeogeogr Palaeoclimatol Palaeoecol 109:1–54

Hoorn C, Wesselingh FP, ter Steege H et al (2010) Amazonia through time: Andean uplift, climate change, landscape evolution, and biodiversity. Science 330:927–931

Hoorn C, Kukla T, Bogotá-Angel G et al (2022) Cyclic sediment deposition by orbital forcing in the Miocene wetland of western Amazonia? New insights from a multidisciplinary approach. Glob Planet Change 210:103717

Jaramillo C, Dilcher DL (2001) Middle Paleogene palynology of Central Colombia, South America: a study of pollen and spores from tropical latitudes. Palaeontographica B 258:87–213

Jaramillo C, Romero I, D'Apolito C et al (2017) Miocene flooding events of western Amazonia. Sci Adv 3:e1601693

Jiménez JA (1984) A hypothesis to explain the reduced distribution of the mangrove *Pelliciera rhizophorae* Tr and Pl. Biotropica 16:304–308

Jones MJ, Sullivan MS, Marsden SJ et al (2001) Correlates of extinction risk of birds from two Indonesian islands. Biol J Linn Soc 73:65–79

Jønsson KA, Irestedt M, Christidis L et al (2014) Evidence of taxon cycles in an Indo-Pacific passerine bird radiation (Aves: Pachycephala). Proc R Soc B 281:20131727

Keesing F, Holt RD, Ostfeld RS (2006) Effects of species diversity on disease risk. Ecol Lett 9:485–498

Keesing F, Belden LK, Daszak P et al (2010) Impacts of biodiversity on the emergence and transmission of infectious diseases. Nature 468:647–652

Liow LH, Stenseth NC (2007) The rise and fall of species: implications for macroevolutionary and macroecological studies. Proc R Soc B 274:2745–2752

Liu C, Sarnat EM, Friedman NR et al (2020) Colonize, radiate, decline: unravelling the dynamics of island community assembly with Fijian trap-jaw ants. Evolution 74:1082–1097

Lo EYY, Duke NC, Sun M (2014) Phylogeographic pattern of *Rhizophora* (Rhizophoraceae) reveals the importance of both vicariance and long-distance oceanic dispersal to modern mangrove distribution. BMC Evol Biol 14:83

López-Martínez N (2009) Time asymetry in the palaeobiogeographic history of species. Bull Soc Geol Française 180:445–455

Lorente MA (1986) Palynology and palynofacies of the Upper Tertiary in Venezuela. Dissert Bot 99:1–222

Losos JB (1992) A critical comparison of the taxon-cycle and character-displacement models for size evolution of *Anolis* lizards in the Lesser Antilles. Copeia 2:279–288

Lu L, Fritsch PW, Matzke NJ et al (2019) Why is fruit color so variable? Phylogenetic analyses reveal relationships between fruit-color evolution, biogeography and diversification. Glob Ecol Biogeogr 28:891–903

MacLean WP, Holt RD (1979) Distributional patterns in St. Croix *Sphaerodactylus* lizards: the taxon cycle in action. Biotropica 11:189–195

Mann P (2021) Gulf of Mexico, Central America, and the Caribbean. In: Alderton D, Elias SA (eds) Encyclopedia of geology. Academic Press, London, pp 47–67

Matos-Maraví P, Matzke NJ, Larabee FJ et al (2017) Taxon cycle predictions supported by model-based inference in Indo-Pacific trap-jaw ants (Hymenoptera: Formicidae: *Odontomachus*). Mol Ecol 27:4090–4107

Muller J, Di Giacomo E, Van Erve AW (1987) A palynological zonation for the cretaceous, tertiary and quaternary of northern South America. Am Assoc Strat Palynol Contr Ser 19:7–76

O'Connell DP, Kelly DJ, Lawless N et al (2019) Diversification of a 'great speciator' in the Wallacea region: differing responses of closely related resident and migratory kingfischer species (Aves: Alceinidae: *Todiramphus*). Ibis 161:806–823

Oliver PM, Brown RM, Kraus F et al (2018) Lizards of the lost arcs: Mid-Cenozoic diversification, persistence and ecological marginalization in the West Pacific. Proc R Soc B 285:20171760

Parenti LR, Ebach MC (2013) Evidence and hypothesis in biogeography. J Biogeogr 40:813–820

Parra FJ, Navarrete RE, di Pasquo MM et al (2020) Neogene palynostratigraphic zonation of the Maranon Basin, Western Amazonia, Peru. Palynology 44:675–695
Pepke ML, Irestedt M, Fjeldså J et al (2019) Reconciling supertramps, great speciators and relict species with the taxon cycle stages of a large island radiation (Aves: Campephaegidae). J Biogeogr 46:1214–1225
Pielou EC (1977) Biogeography. Wiley, New York
Polidoro BA, Carpenter KE, Collins L et al (2010) The loss of species: mangrove extinction risk and geographic areas of global concern. PLoS One 5:e10095
Regali M, Uesugui N, Santos A (1974) Palinologia dos sedimentos meso-cenozóicos do Brasil (I and II). Bol Técnico PETROBRAS 17:177–191 & 263–301
Regali MSP, Uesugui N, Lima EC (1985) Palynostratigraphy and paleoenvironment of the Barreirinhas do Maranhao, Brazil. 7th Congreso Brasileiro de Paleontologia 27, 461–470
Ricklefs RE, Bermingham E (2002) The concept of the taxon cycle in biogeography. Glob Ecol Biogeogr 11:353–361
Ricklefs RE, Cox GC (1972) Taxon cycles in the West Indian avifauna. Am Nat 106:195–219
Romito S, Mann P (2020) Tectonic terrains underlying the present-day Caribbean plate: their tectonic origin, sedimentary thickness, subsidence histories and regional controls on hydrocarbon resources. In: Davidson I, Hull JNF, Pindell J (eds) The basins, orogens, and evolution of the Southern Gulf of Mexico and Northen Caribbean. Geol Soc, London, pp 343–378
Rull V (1998) Evolución de los manglares neotropicales: la crisis del Eoceno. Interciencia 23:355–362
Rull V (2001) A quantitative palynological record from the Early Miocene of western Venezuela, with emphasis on mangroves. Palynology 25:109–126
Rull V (2009) Microrefugia. J Biogeogr 36:481–484
Rull V (2022a) The Caribbean mangroves: an Eocene innovation with no Cretaceous precursors. Earth-Sci Rev 231:104070
Rull V (2022b) Responses of Caribbean mangroves to Quaternary climatic, eustatic and anthropogenic drivers of ecological change: a review. Plan Theory 11:3502
Rull V (2023) Taxon cycles in Neotropical mangroves. Plan Theory 12:244
Sheh SN, Morueta-Holm N, Angert AL (2018) Determinants of geographic range size in plants. New Phytol 226:650–665
Simberloff D, Collins MD (2010) The domain of the dynamic equilibrium theory and assembly rules, with comments on the taxon cycle. In: Losos JB, Ricklefs RE (eds) The theory of Island biogeography revisited. Princeton University Press, Princeton, pp 237–263
Takayama K, Tateishi Y, Kajita T (2021) Global phylogeography of a pantropical mangrove genus *Rhizophora*. Sci Rep 11:7228
Westerhold T, Marwan N, Drury AJ et al (2020) An astronomically dated record of earth's climate and its predictability over the last 66 million years. Science 369:1383–1387
Wijmstra TA (1968) The identity of *Psilatricolporites* and *Pelliciera*. Acta Bot Neerl 17:114–116
Wijmstra TA (1969) Palynology of the Alliance Well. Geol Mijn 48:125–133
Wilson EO (1959) Adaptive shifts and dispersal in a tropical ant fauna. Evolution 13:122–144
Wilson EO (1961) The nature of the taxon cycle in the Melanesian ant fauna. Am Nat 95:169–193
Žliobaitė I, Fortelius M, Stenseth NC (2017) Reconciling taxon senescence with the Red Queen's hypothesis. Nature 552:92–95

Quaternary Shifts and Anthropization 7

Abstract

In the Caribbean region, the oldest Pleistocene mangrove records date from the Last Interglacial (Eemian or MIS 5e), around 130 kyr BP, which is due to the lack of studies on older sediments. During this time, global average temperatures and sea levels were slightly higher than they are today, and mangroves thrived in conditions similar to those found in current locations. Between approximately 70 and 30 kyr BP, during the Last Glaciation, Caribbean temperatures and sea levels were considerably lower (about 5 °C and 100 m, respectively). Mangroves existed in areas far from their current locations, situated on sites now submerged. The present-day configuration of mangroves gradually took shape after the warming of the Early Holocene, starting around 8000 cal yr BP, and the subsequent rise in sea levels, all without significant human influence. Human impact on mangroves became notable in the Mid–Late Holocene. The earliest evidence of human interference with mangroves dates back to around 5500 cla yr BP when Preclassic Maya societies practiced agriculture. They cleared mangroves using fire for cultivating maize and squash, an event known as Mesoamerican Archaic Disturbance (MAD). Throughout the last millennium, ecological changes in mangroves have been primarily driven by episodic droughts associated with the Little Ice Age. Although there is a lack of evidence for mangrove disturbance during known prehistoric cultural crises or European contact, the loss and fragmentation of mangroves due to human activities in recent centuries have been well documented.

This chapter focuses on the response of Caribbean mangroves to natural and human-induced ecological changes during the Quaternary, placing particular emphasis on climatic and sea-level shifts in the Pleistocene and heightened human impact in the

V. Rull, *Origin and Evolution of Caribbean Mangroves*, Ecological Studies 252,
https://doi.org/10.1007/978-3-031-57612-6_7

Holocene. It is crucial to note that at the onset of the Quaternary (2.6 Ma), the closure of the Panama Isthmus had already taken place, interrupting circulation between the Caribbean Sea and the Pacific Ocean. While there is ongoing debate about the exact timing of isthmus closure, consensus suggests it occurred by 2.8 Ma, preceding Quaternary climatic changes (McGirr et al. 2021). Since then, Atlantic circulation has been dominant in the Caribbean Sea. From a biogeographical standpoint, the formation of the Panama Isthmus created a novel terrestrial dispersal route across the American continent but presented an unprecedented barrier for marine organisms, altering the ecological and evolutionary landscape significantly (Baker et al. 2014). For mangroves, this new scenario offered opportunities for both vicariance across the newly established Central American barrier (Nettel and Dodd 2007; Nettel et al. 2008; Cerón-Souza et al. 2010, 2012; Castillo-Cárdenas et al. 2015) and facilitated gene flow along Caribbean and Atlantic coastal currents, crucial dispersal agents for mangroves (Van der Stocken et al. 2019). To avoid circular reasoning when examining mangrove responses to environmental drivers, this chapter primarily relies on vegetation-independent evidence of climatic and eustatic changes. The chapter commences with a summary of Pleistocene and Holocene paleoclimatic and paleoeustatic shifts, accompanied by an overview of human colonization patterns and cultural developments in the Caribbean. The subsequent section underscores the significance of modern-analog studies in paleoecological reconstruction and offers a comprehensive synthesis of available Pleistocene and Holocene mangrove records, focusing on ecological responses to climatic, eustatic, and anthropogenic influences.

7.1 Climatic and Eustatic Changes

7.1.1 Pleistocene

The Pleistocene glacial–interglacial cycles had a broad impact not only on temperate and polar zones but also on tropical regions (Ravelo et al. 2004), including the Caribbean. This region underwent notable temperature fluctuations, alterations in moisture patterns, and sea level changes during this geological epoch (deMenocal et al. 1992; Haug et al. 1998; Yarincik and Murray 2000; Martinez et al. 2007). In the context of this study, the earliest records of Pleistocene Caribbean mangroves identified date back to approximately 130,000 cal yr BP (Mitchell et al. 2001), coinciding with the commencement of the last glacial cycle, spanning from the Last Interglacial, also known as the Eemian or Marine Isotopic Stage (MIS) 5e, to the present (Fig. 7.1). Consequently, our analysis will specifically concentrate on this timeframe, corresponding to the Late Pleistocene. Any potential discovery of unequivocal paleobotanical evidence indicating older Caribbean mangroves in future research will prompt an expansion of the analysis accordingly.

Throughout the most recent glacial cycle, Caribbean SSTs have mirrored global trends. Initially, they followed a pattern akin to the global average, commencing from interglacial (MIS 5e) temperatures higher than those observed today. This was

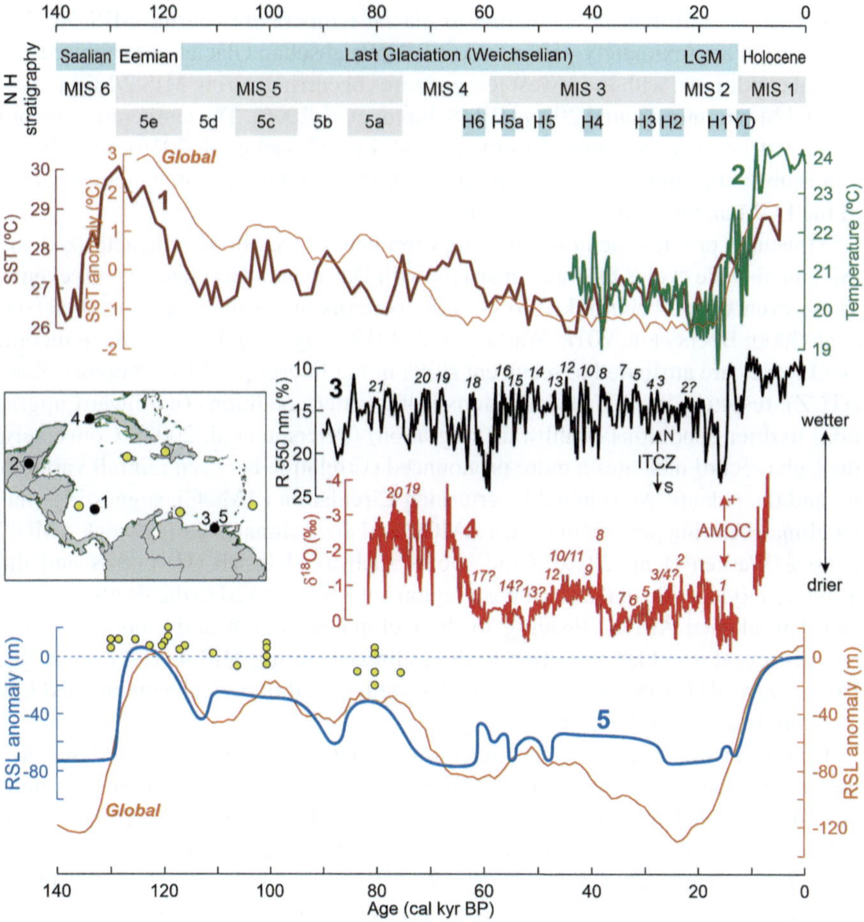

Fig. 7.1 Climatic and sea-level shifts of the last glacial cycle in the Caribbean region. (1) SST (sea surface temperature) reconstruction using the ODP-999A core (Colombian Basin) (Schmidt et al. 2006); (2) continental temperature reconstruction based on the Lake Petén Itzá recod (Guatemala) (Grauel et al. 2016); (3) millennial-scale migrations of the ITCZ (Intertropical Convergence Zone) from reflectance (R) measures in the ODP-1002 core (Cariaco Basin, Venezuela) (Peterson et al. 2000); numbers mark the D-O (Dansgaard-Oeschger) events (Boers et al. 2018); (4) oxygen isotope records from the Santo Tomás cave speleothems (Cuba) (Warken et al. 2019); numbers are D-O events (Boers et al. 2018); (5) Relative sea-level (RSL) reconstruction using seismic stratigraphy studies across the Venezuelan Gulf of Cariaco (blue line) (Van Daele et al. 2011) and several coral terraces (yellow dots) across the Caribbean region (Barbados, Bonaire, Cayman, Colombia, Curaçao, Haiti) (Dumas et al. 2006; Coyne et al. 2007; Hearthy et al. 2007; Creveling et al. 2017; Rubio-Sandoval et al. 2021). Global temperature and sea-level reconstructions from Friedrich and Timmermann (2020) and Spratt and Liesecki (2016), respectively. Northern Hemisphere (NH) glacial stratigraphy based on Martinson et al. (1987), Thompson and Goldstein (2006), and Hughes et al. (2013). AMOC, Atlantic Meridional Overturning Circulation; H1-H6, Heinrich events; ITCZ, Intertropical Convergence Zone; LGM, Last Glacial Maximum; MIS, Marine Isotope Stages; YD, Younger Dryas

succeeded by a further 4 °C decline to glacial temperatures during MIS 5d. The period from approximately 110 to 20 kyr BP (Weichselian Glaciation) captured full glacial conditions, with the lowest temperatures occurring during MIS 2/LGM (Last Glacial Maximum) around 20 kyr BP (Schmidt et al. 2006). The temperature record for terrestrial areas is confined to the past 45 kyr (Grauel et al. 2016), revealing a more substantial amplitude of variation (>5 °C) between the minimum temperatures of the LGM and the current conditions.

Moisture proxies demonstrate high-frequency variations characterized by millennial-scale cycles that are consistent with Dansgaard–Oeschger (D–O) events. These events have been linked to periodic patterns in ocean circulation (Peterson et al. 2000; Boers et al. 2018; Warken et al. 2019) (Fig. 7.1). In the Cariaco record, D–O cycles are attributed to recurrent shifts in the Intertropical Convergence Zone (ITCZ), resulting in cyclical transitions from wetter conditions (northward migration) to drier conditions (southward migration) (Peterson et al. 2000). Conversely, the Cuba record indicates a more pronounced correlation between rainfall variability and the Atlantic Meridional Overturning Circulation (AMOC), suggesting wetter climates during periods of robust AMOC and drier climates during weak AMOC phases (Warken et al. 2019). Cold events such as Heinrich (H) events and the Younger Dryas (YD) in the Caribbean region have been linked to the southern ITCZ and a weakened AMOC, leading to drier climates. Significantly, both moisture records reveal a sudden transition to drier climates during MIS 4 (D–O event 18). However, while Cariaco swiftly returned to wetter conditions, Cuba maintained low moisture levels until the LGM.

Caribbean sea-level reconstructions throughout the last glacial cycle have predominantly focused on coral terraces from MIS 5, with the exception of a continuous curve derived from seismic stratigraphy in the Gulf of Cariaco (Dumas et al. 2006; Coyne et al. 2007; Hearthy et al. 2007; Van Daele et al. 2011; Creveling et al. 2017; Rubio-Sandoval et al. 2021) (Fig. 7.1). The findings consistently indicate sea levels during the Eemian interglacial (MIS 5e) that were 3–20 m higher than present levels. Nevertheless, coral terrace-based inferences consistently yield higher values compared to those obtained from seismic stratigraphy and the globally estimated averages for MIS 5a and MIS 5c. Notably, during the LGM, seismic estimates deviate significantly from the global curve, surpassing the latter by more than 40 m. The subsequent section will delve into a more detailed analysis of the rising trends during the Late Glacial and Holocene periods.

7.1.2 Late Glacial–Holocene

Following the LGM, the well-known Bølling/Allerød–Younger Dryas (B/A–YD) sequence in the Northern Hemisphere, representing interstadial and stadial periods, was observed in both marine and terrestrial paleotemperature reconstructions in the Caribbean (see Fig. 7.2). Notably, the YD cooling was more pronounced and sudden in the marine Cariaco record (Lea et al. 2003; Grauel et al. 2016). The onset of the Holocene featured the Early Holocene Warming (EHW), succeeded by the Holocene

Thermal Maximum (HTM), which is evident in both terrestrial and marine environments. A detailed examination of the last two millennia, as documented by Wurtzel et al. (2013), unveiled climatic fluctuations consisting with the Medieval Climate Anomaly (MCA; 900–1280 CE) and the Little Ice Age (LIA; 1300–1890 CE).

The titanium (Ti) concentration, serving as an indicator for river discharge and continental rainfall, was used to document moisture shifts. According to the Cariaco record, the B/A and EHW–HTM phases were characterized by dominant wet climates, while the YD exhibited arid conditions (Fig. 7.2). Similar to the full glacial conditions, alterations in rainfall on a millennial scale were likely influenced by the latitudinal migration of the ITCZ. The onset of the Late Holocene (4000–3000 cal yr BP) saw century-scale fluctuations in precipitation associated with heightened El Niño activity, giving rise to the term "Cariaco Holocene Instability" (CHI) event used here. This phase featured several extremely dry events comparable in intensity to the YD but of shorter duration (decadal to multidecadal scale). The LIA recorded a distinctive drought event, possibly linked also to ITCZ migration (Haug et al. 2001). Oxygen isotope records from Cuban speleothems over the last millennium suggest that the AMOC played a more significant role than ITCZ migration in the northern Caribbean sector (Fensterer et al. 2012; Tamalavage et al. 2020). Other records also support the previously mentioned climatic shifts, indicating regional environmental trends (Leyden 1985). The Late Holocene presents a more heterogeneous and variable scenario, with increased hydroclimatic spatial variability in Central American (Yucatán), Antillean (Cuba), and northern South American (Cariaco) sites over the last three millennia (Haug et al. 2001; Hodell et al. 2005; Fensterer et al. 2012). Multiyear droughts in the Cariaco record during the last millennium, likely influenced by long-term ITCZ dynamics, have been deemed regionally significant and intense enough to contribute to the cultural collapse of pre-Columbian societies in Central America (Haug et al. 2003).

Over the course of the last 14,000 years, the Caribbean sea levels have exhibited a consistent pattern of increase resembling a logarithmic curve. The highest rates of elevation occurred between 7 and 11 m per millennium during the Late Glacial and the Early Holocene, up to 8200 cal yr BP. Following this period, the rates diminished to less than 2.5 m per millennium in the Middle Holocene (8200–4200 cal yr BP) and further reduced to less than 0.5 m per millennium over the last 4200 years. Throughout the Holocene, the relative sea level (RSL) did not exceed the current height, except in specific locations in northern South America, such as Guyana and Surinam. In these areas, around 7000 cal yr BP, the RSL was approximately 1 m above the present level (Khan et al. 2017).

7.2 Human Settlement

7.2.1 Late Pleistocene

The southern Caribbean coasts of South America, particularly the Taima-Taima site in Venezuela, provide some of the earliest evidence of human presence in the area.

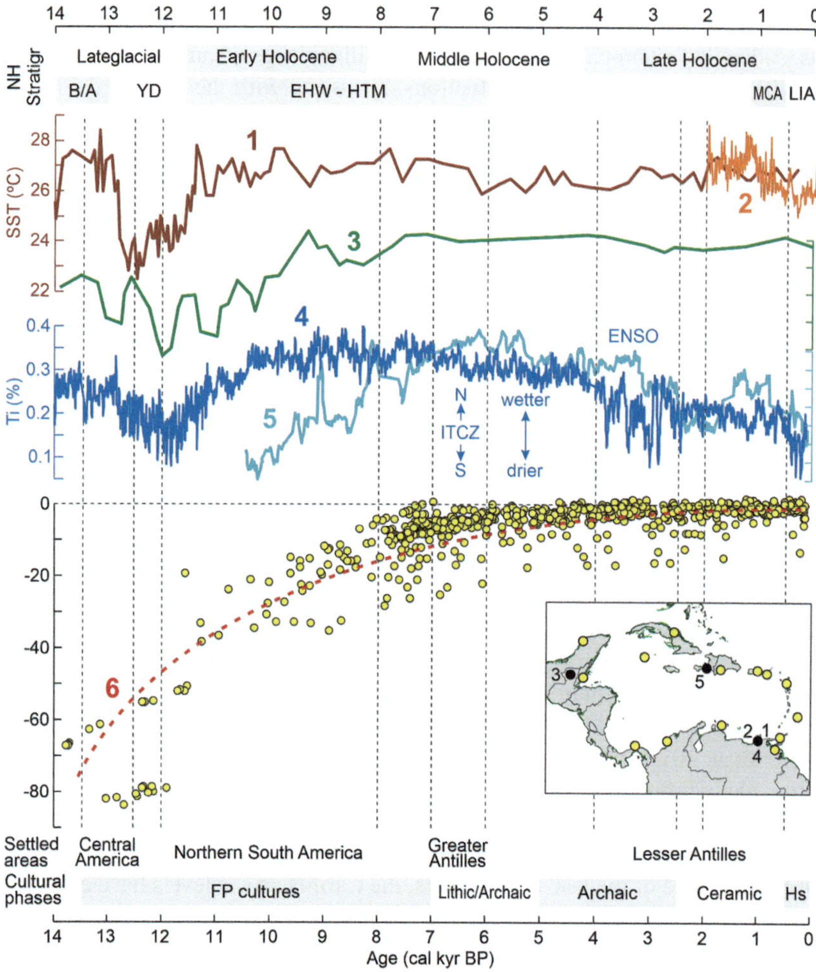

Fig. 7.2 Late Glacial to Holocene climatic, eustatic, and anthropgenic features of the Caribbean region. (1) SST (sea surface temperature) reconstruction from the PL07-39PC record (Cariaco Basin, Venezuela) (Lea et al. 2003); (2) high-resolution reconstruction of the last millennia from PL07-72GGC core (Cariaco Basin, Venezuela) (Wurtzel et al. 2013); (3) continental temperature record using Lake Petén Itzá cores (Guatemala) (Grauel et al. 2016); (4) continental moisture record based on titanium (Ti) concentration, as a proxy for river discharge, in the ODP-1002 record (Cariaco Basin, Venezuela) (Haug et al. 2001); (5) oxygen isotope record from Lake Miragoane (Haiti), as a proxy for moisture (Hodell et al. 1991); (6) sea-level variations using ~500 sites widespread across the Caribbean region (yellow dots in the map) (Khan et al. 2017). B/A, Bølling/Allerød; Cr, ceramic culture; EHW, Early Holocene Warming; Eu, Europeans; FP cultures, Fishtailed-Point cultures; G Antilles, Greater Antilles; Hs, historical; HTM, Holocene Thermal Maximum; ITCZ, Intertropical Convergence Zone; LIA, Little Ice Age; MCA, Medieval Climate Anomaly; YD, Younger Dryas

Excavations at Taima-Taima have revealed artifacts from the Paleoindian period, dating back to 17,300–15,500 cal yr BP. These findings, associated with mastodon remains, suggest early human engagement in megafaunal hunting (Bryan et al. 1978) (Fig. 7.3). This discovery predates both the YD cooling event (see Figs. 7.1 and 7.2) and the Clovis culture, a Paleoindian culture that originated in North America and spread rapidly across South America between 13,000 and 12,500 cal yr BP (Fiedel 2002, 2022). The origin of pre-Clovis cultures and their potential use of the Isthmian land bridge to reach South America remains uncertain (Ranere and Cooke 2021). Some support for this idea comes from Panama, where fragments of undated artifacts, stylistically similar to those found at Taima-Taima, have been discovered (Pearson 2002; Ranere and López 2007; Cooke et al. 2013). However, the majority of evidence from Central America is consistent with the southward expansion of the Clovis culture, dated at 13,400–12,600 cal yr BP (Ranere and Cooke 2021). Therefore, based on the available evidence, it is suggested that the initial human settlements in the Caribbean were along the coasts of South America and Central America. Following this initial wave of expansion, additional settlements occurred in Central and northern South America between the terminal Pleistocene and the Early Holocene, spanning 12,100 to 8200 cal yr BP (Prufer et al. 2019; Ranere and Cooke 2021). These Clovis-derived Paleoindian cultures are characterized by Fishtail Points (FPs) and are commonly known as FP users or FP cultures (Fiedel 2022). Notably, the southeastern Caribbean coasts were inhabited during the Early–Middle Holocene, up to 8000 cal yr BP, taking advantage of lower sea levels that allowed access to the present island of Trinidad by land (Tankersly et al. 2018).

Little information is available regarding the natural resources utilized by Late Pleistocene Clovis and earlier cultures. What is known is that these cultures were primarily foragers and active hunters, targeting now-extinct megafaunal species such as mastodons, toxodons, glyptodons, ground sloths, giant deer, and horses (Ranere and Cooke 2021). There is a proposition that these cultures may have played a significant role in the widespread extinction of megafauna across the continent (Prates and Ivan Perez 2021). Looking at the paleoenvironmental perspective, the expansion of humans into South America via the isthmian bridge coincided with the initial phase of the YD reversal. During this period, climates were notably colder and drier, and the sea level was 60 to 80 m lower than its current position (Fig. 7.2). The decrease in sea level during this time made it easier for people to migrate along the uncovered continental shelf. The development of the FP culture in central and northern South America took place amid the EHW and early HTM periods, characterized by progressively warmer and wetter climates. These conditions likely supported the expansion of human populations.

7.2.2 Holocene

The colonization of the northern and eastern shores of the Caribbean, encompassing the Greater and Lesser Antilles, took place at a later time during the Middle and

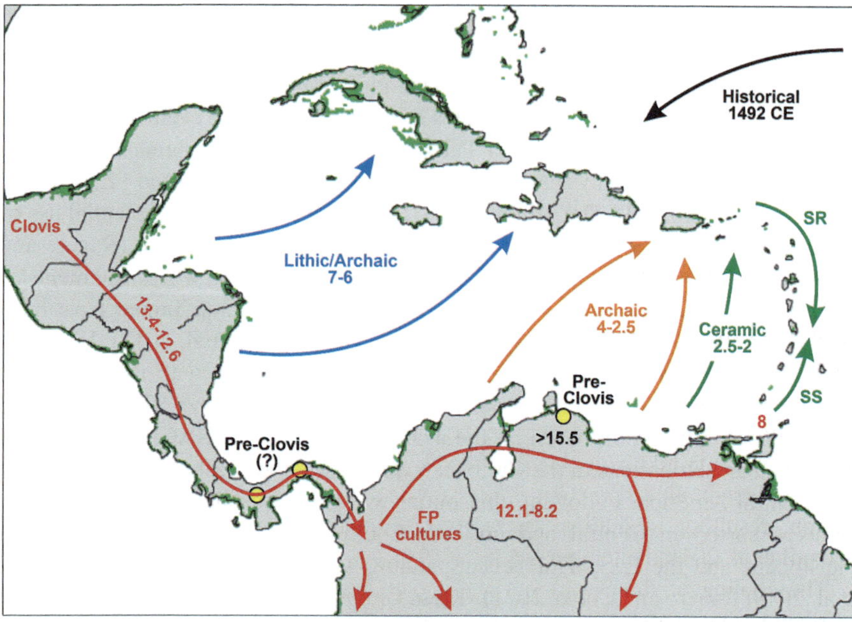

Fig. 7.3 Human settlement of the Caribbean region (composed after Bryan et al. 1978; Morrow and Morrow 1999; Keegan 2000; Pearson 2002, 2017; Fitzpatrick and Keegan 2007; Ranere and López 2007; Cooke et al. 2013; Napolitano et al. 2019; and Nami 2021). Numbers are ages in cal kyr BP. FP cultures, Fishtailed-Point cultures; SR, southern route hypothesis; SS, stepping-stone hypothesis

Late Holocene. Scholars, particularly Keegan and Hofman (2017), traditionally categorize this time span into four main ages: Lithic (7000–5000 cal yr BP), Archaic (5000–2500 cal yr BP), Ceramic (2500–500 cal yr BP), and Historical (1492 CE onward) (Fig. 7.2). Some researchers argue that there is not enough archaeological evidence to differentiate between the Lithic and Archaic ages, suggesting their amalgamation into a single era known as Archaic or Preceramic (Hofman and Antczak 2019; Leepart et al. 2022). In Mesoamerica, which includes southern Mexico and northern Central America, where the Maya civilization flourished, the period from around 4000 to 1700 cal yr BP is termed the Preclassic or Formative period. The Classic and Postclassic periods are delineated for the intervals 1700–1050 cal yr BP and 1050–430 cal yr BP, respectively. It is noteworthy that the climatic phase influenced by the ENSO (CHI), as indicated in the Cariaco record (Fig. 7.2), coincided with the shift from the Archaic to the Mesoamerican Maya Preclassic/Formative period (Neff et al. 2006).

According to the prevailing perspective, the initial settlement of the Greater Antilles is believed to have occurred approximately 7000–6000 cal yr BP, as evidenced by the arrival of Lithic/Archaic peoples in Cuba and Hispaniola, likely originating from Central America (Fitzpatrick and Keegan 2007) (Fig. 7.3). The northern Lesser Antilles and Puerto Rico were inhabited by Archaic peoples from South

America between 4000 and 2500 cal yr BP. Subsequently, between 2500 and 2000 cal yr BP, these islands witnessed the arrival of Ceramic cultures, also of South American origin (Napolitano et al. 2019). However, there are two alternative hypotheses in this scenario. The first, referred to as the stepping-stone (SS) hypothesis, proposes that the migration to the Lesser Antilles originated from South America, moving in a northward direction. Conversely, the second hypothesis, known as the southward route (SR) hypothesis, contends that the Lesser Antilles were inhabited by people migrating from the previously colonized northern Antilles (Napolitano et al. 2019).

Limited information is available regarding subsistence strategies during the Lithic–Archaic ages. Despite earlier beliefs, growing evidence suggests that Archaic Caribbean populations were not solely nomadic foragers relying on coastal resources. Instead, they likely cultivated various plants, including maize in Mesoamerican cultures, established permanent villages, and regularly made and utilized pottery. Anthropogenic fires were a key environmental factor during the Archaic period (Fitzpatrick and Keegan 2007). Environmentally, Preceramic cultures evolved in warm Mid–Late Holocene climates, with a gradual shift toward drier conditions and sea levels slightly below current levels (Fig. 7.2).

More detailed information is accessible regarding ceramic cultures, which thrived between 2500 and 1400 cal yr BP, primarily situated in coastal environments. These cultures employed diverse subsistence strategies, engaging in the hunting of both land and sea animals, fishing, collecting shellfish, and cultivating root crops like manioc (*Manihot esculenta*), a crop introduced to the Caribbean islands by these societies. Additionally, they cultivated a variety of fruits, tubers, and seeds. A significant cultural advancement was marked by the introduction of ceramic technology associated with the cooking and processing of manioc. Given the limited terrestrial fauna, marine foods played a crucial role in their diet, and their fishing methods encompassed the use of hook and line, spears, traps, poison, and nets (Fitzpatrick and Keegan 2007).

By ~1000 cal yr BP, a significant social and environmental transition occurred, probably influenced by a shift from drier to wetter climates and rising sea levels. This change potentially improved agricultural opportunities and the distribution of nearshore environments. Although a shift in pottery styles is documented during this phase, subsistence practices appear to have remained relatively constant (Rouse 1992; Stokes 2005). Nevertheless, there were additions to the diet, such as a wild variety of sweet potato (*Ipomoea* sp.) (Newson and Wing 2004). Generally, during the Ceramic Age, Amerindian groups from South America expanded into nearly every Caribbean island group (excluding the Cayman Islands). This expansion brought about the introduction of new plant and animal species; a decline or extinction of indigenous fish, crabs, rats, and birds on many islands; and an increased reliance on terrestrial and horticulturally important foods, leading to the clearing of forests (Fitzpatrick and Keegan 2007).

7.3 Mangrove Paleoecology

7.3.1 Modern Pollen Sedimentation

Precise interpretation of paleopalynological evidence concerning plant diversity, community composition, and spatial patterns largely rely on taphonomic knowledge. This is commonly addressed through modern analog studies employing the principle of uniformitarianism (Romano 2015). In these studies, modern patterns of pollen sedimentation are compared with the actual features of source plant communities. This approach allows for the derivation of qualitative and/or quantitative relationships, aiding in the inference of past community features from sedimentary pollen records (Jackson and Williams 2004). The roots of modern analog studies in the circum-Caribbean region can be traced back to pioneering palynological surveys conducted approximately six decades ago in the Orinoco delta (Venezuela) and the Guyanan coasts (Muller 1959; Van der Hammen 1963). Notably, the earliest of these studies remains the most extensive, covering a hundred sites across over 100,000 km^2 at a density of one site per 1000 km^2. This comprehensive study provided a detailed depiction of pollen sedimentation patterns throughout the entire Orinoco delta and nearby shallow marine environments (Muller 1959). Subsequently, a dozen sites/areas in the Caribbean region have been investigated for modern pollen analogs with paleoecological objectives (Fig. 7.4).

An initial examination of contemporary sedimentation of mangrove pollen along the Caribbean and Atlantic coasts of northern South America highlighted several key points (Rull 1998):

- In modern pollen assemblages, *Rhizophora* and *Avicennia* emerge as the predominant mangrove trees, with *Laguncularia* and *Conocarpus* generally being scarce or entirely absent.
- *Rhizophora* pollen stands out with its overwhelming abundance, constituting up to 80% in its source area and frequently appearing disproportionately high (overrepresented) compared to the actual prevalence of the parent species. Although this pollen diminishes inland, trace amounts persist beyond the mangrove fringe, and it can be detected up to 80 km offshore in substantial proportions (50%) in shallow marine environments (Muller 1959).
- *Avicennia* pollen, on the other hand, tends to be underrepresented relative to its parent species, seldom surpassing 40% of the pollen assemblage, even in pure *Avicennia* stands. In mixed *Rhizophora–Avicennia* stands, its presence remains below 25%, and its abundance decreases rapidly both inland and offshore (Van der Hammen 1963; Muller 1959).
- Detecting *Conocarpus* pollen poses a particular challenge, and it may be entirely absent even in mangrove stands where this tree dominates (Muller 1959; Rull and Vegas-Vilarrúbia 1999). *Laguncularia* pollen is a rarity.
- The spores of the mangrove fern *Acrostichum* exhibit higher concentrations behind the mangrove fringe, close to the habitat of the species, and are found in lesser quantities in other mangrove soils. Additionally, *Acrostichum* may be

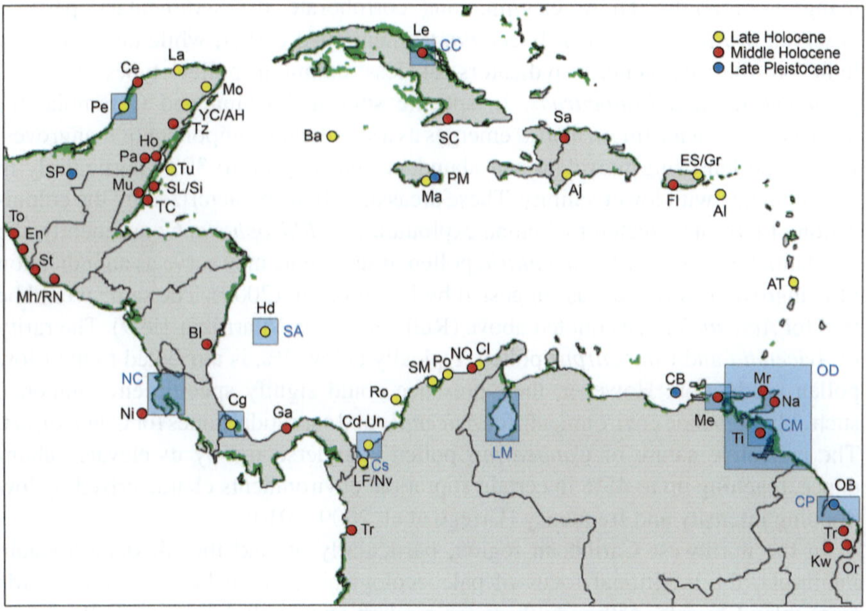

Fig. 7.4 Caribbean sites (dots) and areas (boxes) with Pleistocene–Holocene and modern-analog records, respectively. Dot colors indicate the oldest age of each record. Abbreviations as in Table 2.4 and Fig. 2.8

moderately prevalent in sediments near river mouths, potentially indicating human disturbance (Rull and Vegas-Vilarrúbia 1999).

- These observed patterns are ascribed to interspecific differences in pollen production, transport, sedimentation, and reworking processes, as detailed in the original references. Notably, *Rhizophora*, being wind-pollinated, produces substantial amounts of pollen (DeYoe et al. 2020), while *Avicennia*, *Laguncularia*, and *Conocarpus*, which are insect-pollinated, generate significantly less pollen (Lonard et al. 2017, 2020a, b).

The significance of these sedimentary patterns in interpreting past pollen records is evident. First, the prevalence of *Rhizophora* pollen may overstate the significance of its parent species, whereas the quantity of *Avicennia* pollen may have the opposite meaning. Second, the lack or rarity of *Conocarpus* and *Laguncularia* pollen does not necessarily imply the absence or scarcity of the parent species. Nevertheless, it is stressed that although these overarching observations provide a regional overview of modern sedimentation patterns, interpretations of paleoecology at specific sites should take into account the unique vegetation and deposition characteristics of each location or sedimentary environment.

Subsequent investigations have yielded additional insights, particularly in the context of studies conducted along the coasts of the southwestern Caribbean. These studies have focused on depositional complexes in small bays and lagoons, providing more details about sedimentary sub-environments that hold relevance for local mangrove records. These examinations corroborate the sedimentary patterns observed in *Rhizophora* and *Avicennia*, as mentioned earlier, while also shedding light on the sedimentary indicators of less common pollen types, such as *Laguncularia* and *Conocarpus*. In specific sites in Panama and Colombia, for instance, *Laguncularia racemosa* emerges as a significant component of mangroves, with its pollen being prevalent and abundant, reaching up to 30%, particularly in environments with lower salinity. These areas are often characterized by the colonization of terrains affected by human exploitation of *Rhizophora*. Consequently, the substantial presence of *Laguncularia* pollen in sediments may serve as an indication of mangrove disturbance, as suggested by Urrego et al. (2009). The same would be true for *Acrostichum*, as quoted above (Rull and Vegas-Vilarrúbia 1999). The rarity of *Avicennia* and *Conocarpus* pollen, typically below 2%, is attributed to their low pollen production. However, their presence could signify specific environments, such as hypersaline environments for *Avicennia* and supratidal zones for *Conocarpus*. The indicative nature of *Conocarpus* pollen is underscored by its elevated abundance, reaching up to 45% in certain supratidal environments characterized by low flooding intensity and frequency (Urrego et al. 2009, 2010).

In the northwest Caribbean region, particularly around the Mexican Yucatan Peninsula, the traditional focus of paleoecological research has been on records from inland forests. Although there are modern-analog studies using pollen, very few of them include observations from mangrove environments. A study, involving a few representative mangrove samples, aimed to distinguish modern pollen assemblages from different types of coastal and inland vegetation in this region (Escarraga-Paredes et al. 2014). The findings of the study indicated that sedimentary pollen from typical mangrove tree species characterized this type of vegetation, but the study did not attempt further differentiation within the mangrove community. Similar situations have been observed in other Central American sites, such as those in Costa Rica (Horn 1985; Rodgers and Horn 1996). Yucatan sites that underwent modern analog studies have been used as statistical training sets for developing transfer functions to estimate past precipitation patterns from pollen data. However, no relationships between pollen and vegetation have been established in these studies (Carrillo-Bastos et al. 2013). It is evident that the paleoprecipitation reconstructions derived from these studies cannot be applied to analyze how the vegetation types used for calibration respond to changes in rainfall.

In the northernmost area around the Caribbean, a single study focusing on modern pollen sedimentation in northern Cuba indicates that *Avicennia* pollen is more prominent in modern samples compared to other Caribbean regions. In stands dominated by this tree, it can reach values of 25%, and in hypersaline environments where *A. germinans* coexists with the halophytic saltwort *Batis maritima*, it can reach 65%. In contrast, *Rhizophora* pollen is scarce or absent in this Cuban study, similar to many coasts in northern South America (Davidson 2007). *Laguncularia* and *Conocarpus* pollen are also scarce or absent in this Cuban study area, mirroring observations in various northern South American coasts (Rull 1998).

7.3.2 Paleoecological Records

The extensive literature review conducted for assembling the CARMA dataset has identified 59 sites that contribute valuable information to reconstruct mangrove development during the Late Pleistocene and Holocene in the Caribbean region (Fig. 7.4). Among these sites, only four are related to the Late Pleistocene (>11,700 cal yr BP), and Ogle Bridge in Guyana is the sole location with an Early Holocene sediments (~9570 cal yr BP). The majority of the remaining 55 sites (50) provide records from the Late Holocene (4200–0 cal yr BP), and 32 of them also include sequences from the Middle Holocene (8200–4200 cal yr BP). Notably, there is no apparent geographical pattern in these chronological distributions. The absence of Early Holocene data is particularly significant, coinciding with the expansion of the FP cultures across northern South America (Figs. 7.2 and 7.3). Given that sea levels during this period were 20–80 m below current levels (Fig. 7.2), any sedimentary evidence for Early Holocene mangroves, if present, would likely be submerged under water depths of this magnitude, located far from present-day coasts. Records from the Middle Holocene are scarce before approximately 7000 cal yr BP when sea levels rose to heights similar to the present. The influence of sea level is also evident in the Late Pleistocene, where the three available terrestrial mangrove records correspond to MIS 5, a period when sea levels were either similar to or higher than present levels. In contrast, the marine Cariaco record, derived from sedimentary pollen along the adjacent South American coasts, corresponds to MIS 4–3, a time when sea levels were > 40 m below current levels (Fig. 7.1).

The oldest well-dated Pleistocene evidence of in situ mangroves identified in the CARMA review is associated with the clastic-carbonate Jamaican Port Morant formation at Pera Point (Table 2.4 and Fig. 7.4). These mangroves grew in a coastal lagoonal environment during the Last Interglacial period. Within the sedimentary facies VII of this formation, there is fossiliferous sandstone containing oysters with attachment scars of mangrove roots, suggesting deltaic mangrove environments with typical marine salinities. An electron spin resonance (ESR) date on coral material from this facies indicates an age of 132 ± 7 kyr BP, coinciding with the earliest highstand of MIS 5e (Mitchell et al. 2001). At that time, sea levels were significantly higher than current levels, and SSTs were around 4 °C higher than today (Fig. 7.1). Despite the limitation in obtaining more precise taxonomic and environmental information, it is clear that these Late Pleistocene Caribbean mangrove ecosystems flourished in climates warmer than the present, characterized by higher sea levels.

Another Late Pleistocene record of comparable age and environmental conditions has been documented in interior Mexico, specifically at San Pedro in the southern Yucatan Peninsula (Table 2.4 and Fig. 7.4). In this instance, a mangrove community dominated by *R. mangle* was identified inland, situated far from coastal areas and saline waters (Aburto-Oropeza et al. 2021). However, it is known that the wide environmental adaptability of this species allows it to thrive in calcium-rich freshwater environments (Koch and Snedaker 1997; Hanagata et al. 1999). Molecular phylogenetic studies, employing a molecular clock of 3.16×10^{-9}

mutations per nucleotide per generation, estimated an age range of 5–273 kyr BP (with an average of 98 kyr BP) for the divergence of the San Pedro mangroves and their closest sister coastal lineage, located approximately 200 km north in SW Yucatan (Aburto-Oropeza et al. 2021). It was proposed that these inland mangroves could be remnants of Eemian (MIS 5e) coastal mangroves, dating back to a time when sea levels were around San Pedro, positioned 10 m above the current sea levels. This hypothesis is in agreement with the aforementioned Jamaican record and the overall scenario in the Caribbean region, as outlined in Fig. 7.1.

7.3.2.1 Last Glaciation

A third Late Pleistocene mangrove record from the Venezuelan Cariaco Basin corresponds to the last glacial period, within the MIS 4/MIS 3 interval, spanning approximately 70 to 30 kyr BP. This time frame encompasses Heinrich events H6 to H3 and D–O events 20 to 6 (Fig. 7.1). The Cariaco Basin constitutes a crucial site for understanding past climatic conditions in the Caribbean region, which has been extensively documented in studies by Haug et al. (1998, 2001), Peterson et al. (2000), and Yarincik and Murray (2000), among others. The palynological data from the Cariaco Basin were obtained from a marine core (MD03-2622; Fig. 7.1) located at a water depth of nearly 900 m. This core serves as a proxy to monitor changes in the environment and vegetation along the neighboring northern South American coasts, the primary source of the pollen recovered in the record. Notably, during cold and dry Heinrich events, there were sudden expansions in saltmarsh pollen, indicating a typical chenopod–grass–sedge community succession (Fig. 7.5). This phenomenon, referred to as the Heinrich-event vegetation mode, was accompanied by distinct decreases in forest assemblages and overall pollen input from the continent (González and Dupont 2009). Furthermore, warm and wet D-O peaks recorded an interstadial mode characterized by opposite pollen trends, specifically an expansion of semideciduous and evergreen forests originating from the continental coastal range (González et al. 2009).

The only true-mangrove elements documented in this Caricao record were *Rhizophora* and *Acrostichum*, each accounting for less than 4% of pollen assemblages. While the original reference (González et al. 2009) categorizes these mangrove constituents as "littoral vegetation," the analysis presented here separates them to explore their individual trends more comprehensively. Notably, during H-events (Fig. 7.5), *Rhizophora* and *Acrostichum* consistently showed lower values, indicating that these abrupt cold/dry reversals, combined with drops in sea level, were unfavorable for mangrove development in the littoral fringe. Furthermore, *Rhizophora* exhibited rapid but minor increases immediately after each H-event, suggesting a swift response to post-H warmings, followed by subsequent declines. The significant relationship between mangrove elements and D-O warming supports this interpretation. Therefore, it can be concluded that mangrove communities underwent modest expansions during warm/wet/highstand conditions and contractions under opposing circumstances.

The most recent Pleistocene mangrove record available, the Ogle Bridge record (Table 2.4; Fig. 7.3), originates from British Guiana, currently known as Guyana.

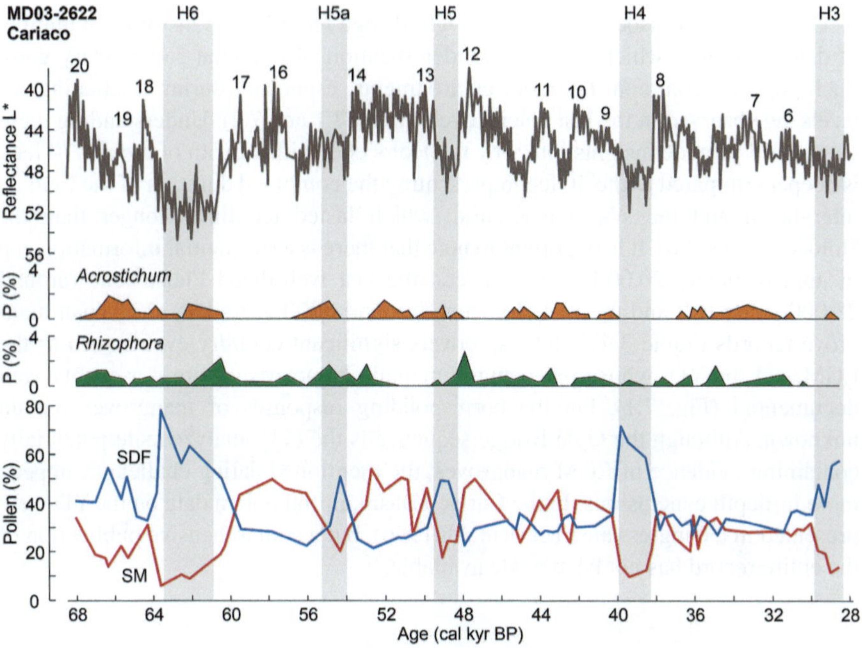

Fig. 7.5 Simplified pollen record from the Cariaco core MD03-2622 highlighting shifts in true-mangrove elements such as *Rhizophora* and *Acrostichum*. The reflectance, semideciduous forest (SDF), and salt marsh (SM) curves are provided as a reference. H6 to H3 are Heinrich events, and numbers are D–O events. Composed from González and Dupont (2009) and González et al. (2009)

This record is featured by a sequence with poor chronological control, assumed to cover the Last Interglacial, the Last Glaciation, and the Holocene (Fig. 7.6). The sequence is bracketted by only two dates, one exceeding the ^{14}C range and the other corresponding to the Early Holocene (~9570 cal yr BP). This latter date is the sole instance of Early Holocene age of the CARMA dataset. The chronological interpretation of the Ogle Bridge sequence heavily relies on pollen stratigraphy (Van der Hammen 1963); therefore, attempts to analyze vegetation responses to climatic shifts based on this sequence could lead to circularity. It is important to note that the identification of the Last Glaciation in the Ogle Bridge record is based on the notable decline of mangroves, giving way to grass savannas, as indicated by pollen percentages in the interval between approximately 74 and 84 feet. Therefore, claiming that mangroves significantly reduced and savannas expanded dramatically due to cold glacial conditions would constitute a logical fallacy. Nevertheless, support for this interpretation can be found in previously discussed examples from Jamaica, Mexico, and the Venezuelan Cariaco Basin. In these records, warm/wet climates and high sea levels consistently correlate with mangrove expansions, while cold/dry conditions and low sea levels coincide with mangrove declines in favor of other coastal communities.

One limitation associated with the Ogle Bridge record is the insufficient number of dated samples, which hinders the identification of potential sedimentary gaps. Such gaps are common in coastal environments, especially during fluctuating sea levels, as observed in the last glacial cycle (Figs. 7.1 and 7.2). Understanding these gaps could provide insights into why the Holocene, with a depth of around 70 feet, is deeper compared to the 30 feet representing the combined duration of the Eemian interglacial and the Würm glaciation, which lasted ten times longer than the Holocene (Fig. 7.6). It is important to note that there is a substantial information gap of approximately 20,000 years between the last well-dated Pleistocene (around 28,000 years BP) and the first Holocene (around 8000 years BP) Caribbean mangrove records (Table 2.4). This gap covers significant cool/dry events such as the LGM and the YD, whose paleoenvironmental characteristics are reasonably well documented (Fig. 7.1), but the corresponding responses of mangroves remain unknown. Although the Ogle Bridge sequence is the only analyzed site potentially containing evidence of LGM mangroves, the mentioned dating challenges impede more in-depth conclusions. In the Cariaco Basin, initial pollen data on the YD were presented in a congress abstract (Rinaldi 1996), but a comprehensive publication of the entire record has not been made available.

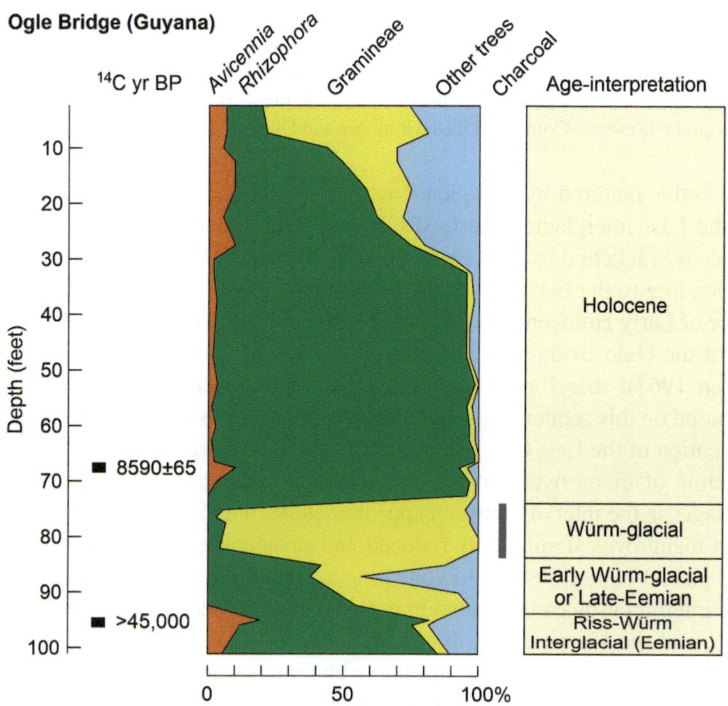

Fig. 7.6 Summary pollen diagram from the Ogle Bridge section (Guyana) and chronological interpretation, according to the original. Other trees include *Alnus*, *Ilex*, *Symphonia*, *Virola*, *Mauritia*, and other palms. Redrawn from Van der Hammen (1963)

7.3.3 Holocene

As previously mentioned, the records of mangroves during the Holocene commence in the Middle Holocene, by 8000 cal yr BP. During this period, temperatures were comparable to the present day, and sea levels were slightly below their current position (Fig. 7.2). This timeframe coincided with the presence of FP cultures along the continental Caribbean coasts and occurred shortly before the Lithic/Archaic peoples settled in the Greater Antilles (Fig. 7.3). The most comprehensive Holocene mangrove record is found at a site in Central America (SL, Belize), covering the last ~8000 years (Monacci et al. 2009). Other relatively complete and well-dated records spanning the past ~7000 years are from sites Si (Monacci et al. 2011) in Belize and Mr in Trinidad (Ramcharan and McAndrews 2006) (Table 2.4; Fig. 7.7). However, these sequences have limitations, including a substantial sedimentary gap between approximately 6000 and 1000 cal yr BP for SL and Si, and insufficient time resolution for the Mr pollen record (Fig. 7.8).

A cursory examination of the diagrams presented in Fig. 7.8 quickly highlights significant spatial variations in the composition of Holocene mangroves. This differences are apparent even in closely situated sites like SL and Si, separated by approximately 20 km. The compositional differences observed in local mangroves persist to the present day. Specifically, mangroves in SL are predominantly characterized by *Rhizophora*, with the presence of *Avicennia* and *Laguncularia* (Monacci et al. 2009). In contrast, the mangrove forests in Si, located more inland, show a

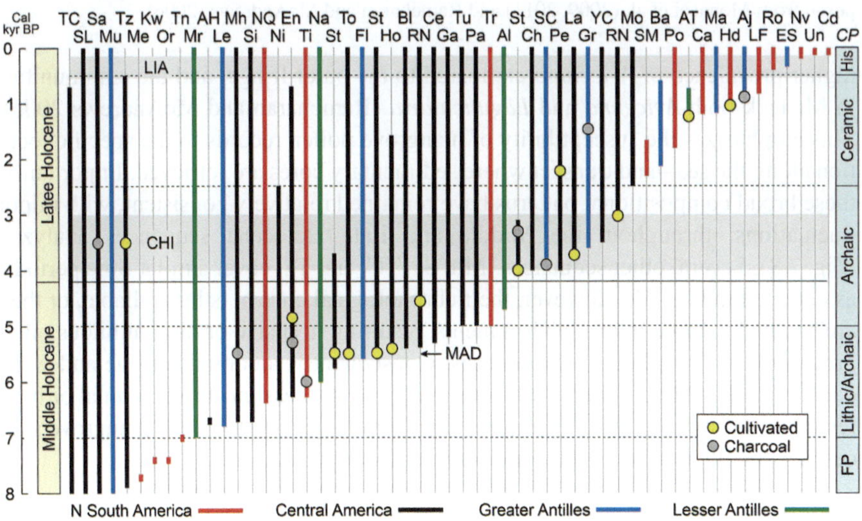

Fig. 7.7 Holocene pollen records from the Caribbean region sorted chronologically. Line colors indicate subregions. First appearances of charcoal and pollen from cultivated plants are indicated by colored dots. Data and abbreviations from Table 2.4 (see Fig. 7.4 for location). CHI, Cariaco Holocene Instability event; *CP*, cultural phases; His, historical period; MAD, Mesoamerican Archaic Disturbance event

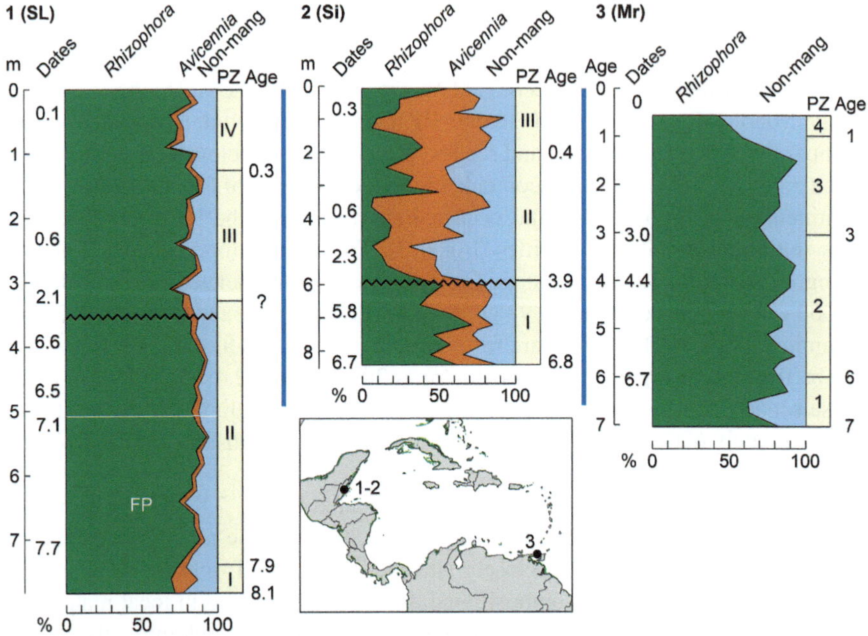

Fig. 7.8 Summary pollen diagrams for the sites with more complete Holocene records in the Caribbean region. Dates, in cal kyr BP, have been interpolated or extrapolated from age-depth models in the original references. The approximate boundaries of the sedimentary gaps in diagrams SL and Si are indicated. FP, Fishtailed Point cultures; PZ, pollen zones. Redrawn and composed from Monacci et al. (2009, 2011) and Ramcharan and McAndrews (2006)

higher prevalence of *Avicennia* compared to the outer fringe, and the communities in Mr lack both *Avicenna* and *Laguncularia* (Ramcharan and McAndrews 2006). This emphasizes the susceptibility of mangrove pollen records to the specific location along the sea–land transect where sedimentary cores are retrieved. Importantly, these broad compositional patterns have remained relatively consistent, with minor fluctuations, throughout the Middle and Late Holocene sequences analyzed (Fig. 7.8). Several other sequences (Mu, Sa, TC, and Tz) cover similar time periods, as depicted in Fig. 7.7. However, their chronological control is less reliable, or they are situated in more inland environments, resulting in sparse or discontinuous mangrove pollen throughout the sequences. Consequently, a comprehensive and well-dated record of Middle–Late Holocene mangrove succession at a centennial or finer resolution is not available. Therefore, any examination of Holocene mangrove succession should depend on the composition of shorter records.

7.3.3.1 Middle Holocene

The only well-documented record providing details about the FP cultural phase (>7000 cal yr BP) is SL (Fig. 7.8). This time period encompasses Pollen Zone I (PZ-I) and the lower segment of PZ-II (Monacci et al. 2009). PZ-I (8100–7900 cal yr BP) signifies the onset of mangrove presence when rising sea levels reached the location, suggesting that by ~8000 cal yr BP, the sea level was approximately 8 m below its current level. This finding is consistent with regional reconstructions (Fig. 7.2). Similar evidence of early mangrove colonization along Caribbean shores due to rising sea levels is found at site TC in Belize (MacIntyre et al. 2004) and several locations in northern South America, such as Me in Venezuela (Rull et al. 1999) and Kw, Or, and Tn in Guyana (Van der Hammen 1963). In the SL record, the initial mangrove communities were dominated by *Rhizophora*, with *Avicennia* present and *Laguncularia* absent. Mangroves fully established themselves at the site during PZ-II (7900–4200 cal yr BP), with *Rhizophora* slightly expanding at the expense of *Avicennia*, which underwent a modest yet significant decline and never regained its former abundance observed in PZ-I. Although not explicitly stated in the original reference, this pattern may suggest a noteworthy role of *Avicennia* in the initial colonization stages, supported by the well-known pioneering nature of *A. germinans* in the upper intertidal fringe (Lonard et al. 2017). Regarding climate, the regional peak in HTM precipitation, characteristic of the Caribbean region and associated with the northward migration of the ITCZ (Fig. 7.2), was also evident in the SL section using independent oxygen isotopic data (Monacci et al. 2009). Consequently, the complete development of mangroves during PZ-II was likely facilitated by highstand conditions and wet climates.

A well-documented account of how mangroves have responded to rising sea levels over the past 7000 years comes from the Ti site in the Orinoco Delta. Presently situated approximately 20 km inland, this site no longer supports mangrove vegetation; instead, it features freshwater stands dominated by *Mauritia* palms (Montoya et al. 2019). However, the scenario was different prior to 5200 cal yr BP when *Rhizophora* was present at the site closer to the intertidal zone. This occurred during a period when sea levels were lower than they are today, as indicated by independent evidence from nearby sites (Rull et al. 1999). This stands in stark contrast to other locations in the Caribbean where the presence of Holocene mangroves is typically linked to rising, not falling, sea levels. In those instances, it is generally believed that mangroves follow the vertically shifting intertidal zone controlled by eustatic factors. However, at the Ti site, the rise in sea levels appears to be more closely associated with lateral movements that displaced the delta front coasts seaward, resulting in a shift away from the coring locality. This phenomenon could result from the enormous sediment influx from the extensive watershed of the Orinoco River, leading to rapid coastal progradation. Consequently, lateral shifts in intertidal environments at the Ti site are significantly faster than vertical sea-level displacements (Vegas-Vilarrúbia et al. 2015).

Until 6000–5500 cal yr BP, changes in the climate, particularly in moisture levels and eustatic shifts, were the main factors influencing mangrove ecosystems. However, after this period, human activity emerged as a significant contributor. The impact of humans on Caribbean mangroves is typically assessed through sources independent of proxies used to track vegetation changes, such as the presence of pollen or phytoliths from cultivated plants and notable increases in sedimentary charcoal (Neff et al. 2006). Maize (*Zea mays*) is particularly noteworthy as the primary cultivated plant in the Caribbean during the Holocene, with evidence suggesting its use by Mesoamerican lowland peoples dating back to >5000 cal yr BP (Pope et al. 2001). Another relevant cultivated plant, squash (*Cucurbita* spp.), is often inferred from the presence of Cucurbitaceae pollen (Xia et al. 2021). In the Neotropics, the consistent presence of significant amounts of charcoal in Late Pleistocene and Holocene sediments has proven to be a robust indicator of human presence (Bush et al. 2007; Rull et al. 2015). Unlike pollen from cultivated plants, which has been routinely identified since the early stages of Holocene pollen analysis (Faegri et al. 1989), the use of charcoal records as a fire proxy is a more recent and less common practice. Consequently, the absence of charcoal in paleoecological records does not necessarily indicate the absence of anthropogenic fires at a specific site and could be due to the lack of the corresponding charcoal studies.

In the regions being examined, the earliest recorded occurrences of pollen from cultivated plants can be traced back to approximately 5500–5600 (cal yr BP at sites Ho, To, and St in Mesoamerica (Neff et al. 2006; Aragón-Moreno et al. 2018; Bocanegra-Ramírez et al. 2019). These sites hold some of the oldest documented records in the Caribbean region. Similarly, when reviewing charcoal records indicating human presence, the oldest discoveries, indicative of human activity, also date back to similar periods and originate from the same region (sites St and Mh), except for the Ti locality in the Orinoco Delta, which boasts the earliest record (~6000 cal yr BP) to date (Montoya et al. 2019). Together, these findings imply an initial interaction of humans with mangroves between 5600 and 4500 cal yr BP, during the shift from the Lithic to the Archaic ages, if these categorizations are considered valid. If the Lithic Age is not considered to be valid, this timeframe, identified as the Mesoamerican Archaic Disturbance (MAD) event in this book (Fig. 7.7), occurred around the middle of the Archaic Age. In the period just preceding this event, between 6000 and 5400 cal yr BP, a significant increase in charcoal is documented in the SE Caribbean region (Ti site in the Orinoco Delta). This surge is attributed to the arrival of the first human colonizers (Montoya et al. 2019) and represents the earliest proof of human presence during the Holocene. The precise origins of these early human settlers remain uncertain, but considering the broader context of human occupation in northern South America (Fig. 7.3), there is a possibility that they were associated with FP cultures.

Well-dated MAD records, covering a span of less than a century, are discoverable in various locations, including sites En and St in southern Mexico and Guatemala, respectively (Fig. 7.9). Concerning site En, the MAD event is thoroughly documented from around 5400 to 4500 cal yr BP. Throughout this timeframe, there was a sudden increase in disturbance indicators, particularly the presence of Chenopodiaceae/Amarantheceae, coinciding with a significant decrease in *Rhizophora* and an increase in *Laguncularia*, known for actively colonizing disturbed *Rhizophora* mangroves (Urrego et al. 2009). This abrupt change in vegetation was accompanied by the initial appearance of Cucurbitaceae pollen and the initiation of continuous charcoal records, although, unfortunately, only the presence or absence of charcoal was recorded in this particular case. The overall scenario strongly suggests landscape anthropization associated with the commencement of agricultural practices (Joo-Chang et al. 2015).

The MAD event is also apparent in the St record (core SIP001) from 5500 to 4600 cal yr BP. During this timeframe, secondary colonizers like aquatic taxa (e.g., Cyperaceae and *Typha*) emerge after mangrove clearing, and maize is identified as the first cultivated plant (Fig. 7.9). It would be interesting to know whether and how mangrove soils were desalinized for maize cultivation. In this case, quantitative charcoal estimates (particles/mL of sediment) provide a more precise evaluation of the local anthropogenic fire incidence compared to the background signal. Similar patterns are observed in neighboring sites (RN, Mh, and other St cores; Table 2.4 and Fig. 7.4) within comparable time intervals (Neff et al. 2006). This suggests that the MAD event extended beyond a local context, even though current evidence indicates its limitation to the Guatemala-Mexico border area (Fig. 7.10). Notably, records from nearby regions, like the Yucatán site Ho, do not indicate a similar mangrove disturbance despite the consistent presence of substantial maize pollen throughout the entire section (5600–1700 cal yr BP) (Aragón-Moreno et al. 2018). It is important to recall that the individuals involved in these activities were not solely foragers collecting shellfish and coastal resources; they also practiced slash-and-burn farming (Kennett et al. 2010). These cultures likely contributed to the

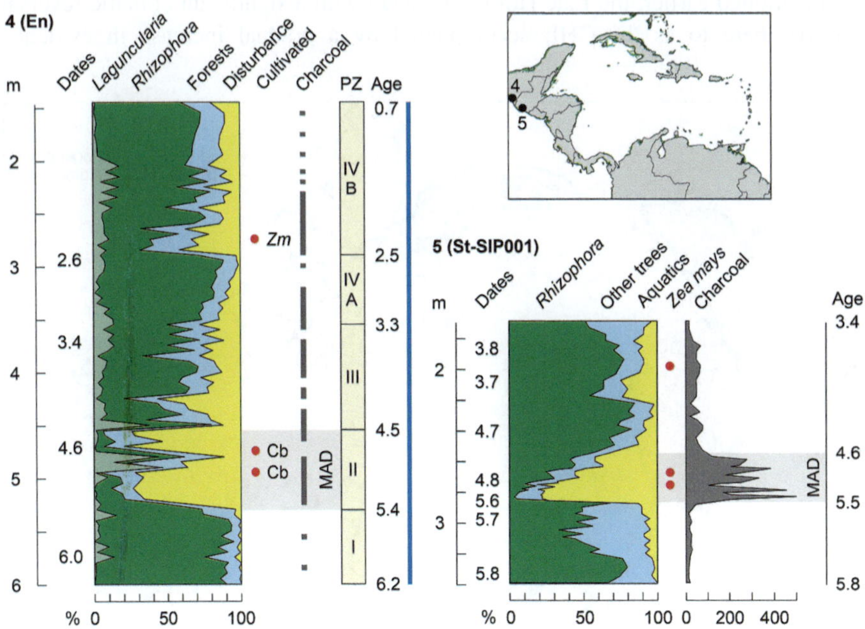

Fig. 7.9 Summary subcentennial-scale pollen records from Mesoamerica including the MAD (Mesoamerican Archaic Disturbance) event. Red dots indicate the occurrence of pollen from cultivated plants (Cb, Cucurbitaceae; *Zm*, *Zea mays*). Dates, in cal kyr BP, have been interpolated or extrapolated from age-depth models in the original references. Charcoal is represented in concentration units (particles/mL of sediment). PZ, pollen zones. Redrawn and composed from Neff et al. (2006) and Joo-Chang et al. (2015)

population that initially inhabited the Greater Antilles (refer to Fig. 7.3) (Pearson 2017).

In other areas, human activities had a relatively minor impact compared to the significant influence of climatic factors, specifically changes in precipitation, on mangrove dynamics. An illustration of this is seen at the Yucatan Ho site discussed earlier, where independent geochemical paleoclimatic evidence indicates that during the MAD event, the ITCZ shifted northward, resulting in increased precipitation. Consequently, this facilitated the expansion of inland rainforests and freshwater wetlands, coinciding with a period of minimal mangrove cover (Joo-Chang et al. 2015). Natural events, such as hurricanes and landfalls, exhibited a relatively calm period between 5400 and 4900 cal yr BP. This coincides with lower sea levels and the prevalence of freshwater wetlands, leading to a reduction in the extent of mangroves along the Nicaraguan coasts (site Bl). Notably, this decline occurred without apparent human influence (McCloskey and Liu 2012). Overall, there is limited or no evidence of human disturbance during the Mid-Holocene outside of Mesoamerica. Instead, predominant factors in the Caribbean region were natural environmental elements, particularly eustatic shifts and fluctuations in moisture.

7.3.3.2 Late Holocene

As mentioned earlier, the Late Holocene began with a significant climatic reversal referred here to as the CHI, accompanied by a gradual increase in evidence

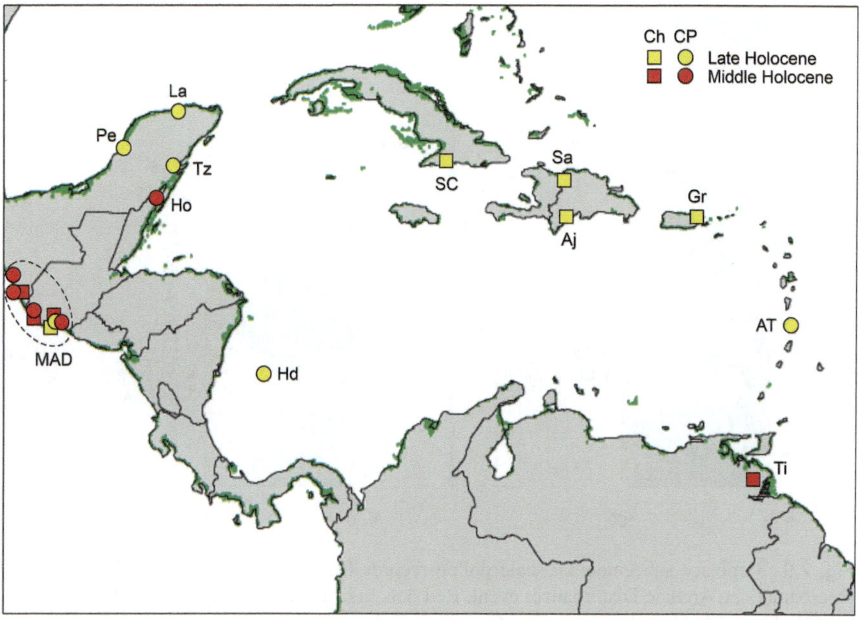

Fig. 7.10 First appearances of charcoal and pollen from cultivated plants in the Caribbean mangrove records. Raw data from Table 2.4 (see also Fig. 7.7). Ch, charcoal; CP, cultivated plants; MAD, Mesoamerican Archaic Disturbance event

indicating human impact across different locations and over time (see Fig. 7.7). Over the last 4200 years, there has been a notable rise in pollen from cultivated plants and charcoal, suggesting human disturbance, spreading throughout the entire Caribbean region, including the previously inhabited Antilles (Fig. 7.10). It is important to highlight the absence of such evidence along the southern and western Caribbean coasts. However, whether this absence is due to the lack of human influence on mangroves or the absence of charcoal studies (problems in identifying maize and/or squash pollen are unlikely) remains uncertain. Therefore, it is crucial to consider both natural and anthropogenic factors, along with their potential feedbacks and synergies, as drivers of ecological change in mangroves across the entire Caribbean region during the Late Holocene. The impacts of climatic, eustatic, and anthropogenic factors display spatial heterogeneity. For example, during the CHI phase (4000–3000 cal yr BP), regional moisture shifts have been identified as the primary drivers of mangrove change in a Yucatan locality (Tu), while human disturbance has been suggested as the main cause of mangrove change in the Mesoamerican site St (Neff et al. 2006; Wooler et al. 2009).

Pollen records from Tu and St, although of low resolution, offer a well bracketted chronological framework (Fig. 7.11). Both locations show a significant decrease in mangroves at the onset of the CHI reversal, replaced by terrestrial and aquatic herbs. Despite these similar trends, distinct factors influenced the shifts in vegetation at each site. In Tu, the decline of *Rhizophora* mangroves (Pollen Zone III, 3900–3300 cal yr BP) oincided with elevated $\delta^{13}C$ values, indicating increased salinity due to higher evaporation rates in drier climates. This is consistent with local climate reconstructions (Wooler et al. 2009) and the broader CHI shift observed in the Cariaco Basin (Haug et al. 2001). The absence of indicators of human disturbance, such as pollen from cultivated plants or increased charcoal, supports the interpretation of climate-driven mangrove decline in this locality. In contrast, at the St site, the reduction of *Rhizophora* mangroves coincided with a notable charcoal peak and intensified maize cultivation (Fig. 7.11). This suggests a second Mesoamerican anthropization pulse at the beginning of the Preclassic Maya period (Neff et al. 2006). The chronological coincidence of this anthropization event with the regional CHI drying trend raises the possibility of climate–human feedbacks that could have amplified mangrove decline at St.

In other Caribbean sectors, mangrove communities have remained relatively stable despite fluctuations in local climate and eustatic conditions. A notable instance of this constancy is observed in the Ga site in Panama, where mangroves have maintained consistency over the past ~5000 years. Throughout the Holocene, the inferred cover and composition of these mangroves, including species like *Rhizophora*, *Avicennia*, and *Acrostichum*, have remained fairly constant, with only minor variations. This ability to endure is attributed to the complex interplay of factors such as climate, sea-level changes, and lateral coastal migration, as proposed by Castañeda-Posadas et al. (2022). In the early phase of the Ga record (5200–5000 cal yr BP), mangroves migrated inland due to a gradual rise in sea levels, which were then 4–5 meters below their current levels. Subsequently (5000–3100 cal yr BP), there was a period of mangrove recession, likely influenced by intensified landward

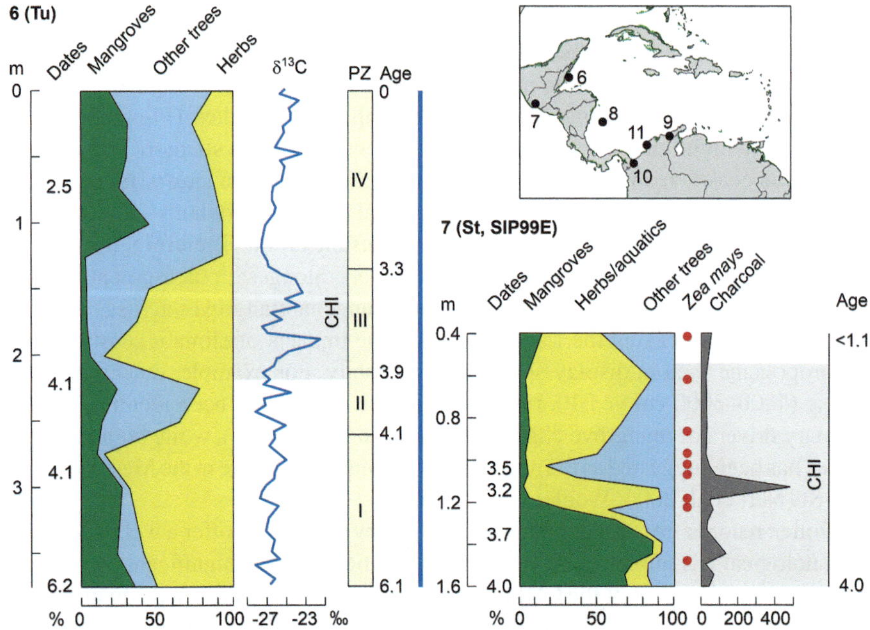

Fig. 7.11 Summary pollen diagrams for the Mesoamerican Late Holocene including the time interval corresponding to the CHI reversal. Red dots mark the presence of pollen from cultivated plants. Dates, in cal kyr BP, have been interpolated or extrapolated from age-depth models in the original references. Charcoal is represented in concentration units (particles/mL of sediment). PZ, pollen zones. Redrawn and composed from Neff et al. (2006) and Wooler et al. (2009)

migration, possibly triggered by drier climates such as the CHI event. This event, prompted by the combined effects of ITCZ migration and ENSO intensification, is supported by lower negative $\delta^{13}C$ values and higher C/N ratios, indicating a prevalence of marine organic sedimentation. After a phase of maximum sea levels and minimal terrigenous sediment input (3000–1800 cal yr BP), there was a notable increase in mangrove pollen between 1800 and 700 cal yr BP. This growth is attributed to coastal progradation resulting from increased terrigenous input, probably facilitated by wetter climates. Importantly, throughout this chronological sequence, there are no indicators suggesting anthropogenic disturbance.

These observations imply that the sustained presence of consistent mangrove cover and community composition does not necessarily indicate a stable environment. It serves as a warning against relying solely on vegetation indicators to reconstruct environmental changes when studying how plant communities react to climate and sea-level influences. The results from this study, as well as others in the Orinoco Delta and NW Colombian coasts (sites Ti and NQ), highlight that Holocene mangrove records are significantly shaped by the interaction between terrigenous input—largely influenced by the size of incoming river systems and continental precipitation—and sea-level rise. In periods of arid climate, the predominance of sea-level rise triggers the landward migration of mangrove communities. Conversely, during wetter phases, sedimentation can compensate for or surpass eustatic rise, resulting in a seaward migration along the coast. This underscores the importance of careful selection of evidence when documenting mangrove responses to climate and sea-level changes, advocating for the use of pollen-independent proxies to prevent circular reasoning. Specifically, it is advised to avoid inferring sea-level trends from mangrove dynamics and analyzing mangrove responses to inferred eustatic fluctuations using the same evidence.

Over the past 3000 years, there has been no evidence of substantial environmental shifts in the Caribbean region, apart from the previously mentioned prolonged YD drying reversal (Fig. 7.2). In contrast, intensified anthropogenic disturbance has become a significant force driving changes in mangrove ecosystems across the region (Fig. 7.10). An illustration of human influence is evident in the En site discussed earlier, dating back to 2500 cal yr BP. At that time, the reappearance of *Zea mays* pollen in the diagram and continuous charcoal records coincided with a sudden decline in mangroves and a notable increase in disturbed vegetation (Fig. 7.9). Similar patterns were observed in certain St sections during the Preclassic Maya period (Neff et al. in 2006).

Other records of mangrove decline over the past three millennia have been explained in the context of either climate fluctuations or interactions between climate and human activities. For instance, a significant decrease in *Rhizophora* mangroves at the Gr site in Puerto Rico, spanning from 1550 to 575 cal yr BP, was linked to a period of heightened tropical storm activity (Lane et al. 2013). This decline in mangroves also coincided with a notable increase in charcoal deposits, suggesting human presence. Although there is a lack of pollen from cultivated plants, this interpretation was strengthened by the discovery of human-made artifacts dating back to approximately 2000 cal yr BP (Ceramic age) near the coring site. Similarly, in Hispaniola, a reduction in *Rhizophora* accompanied by a significant rise in Amaranthaceae, including halophytic species and those indicative of arid conditions and disturbance, was observed since 2500 cal yr BP. This phenomenon was interpreted as a response to drier climatic conditions (Caffrey et al. 2015). This instance was conceived to be a response to more arid climate conditions (Caffrey et al. 2015). The elevated charcoal abundance by 1500 cal yr BP in the same record was ascribed to heightened fire activity triggered by dry climates, rather than human actions, supported by the absence of signs of human settlement in the area prior to European contact. In analogous records from northern South America with comparable ages, the decrease in mangrove presence was also connected to arid climates, and subsequent expansions were linked to humid conditions, as exemplified by the Colombian site NQ (Urrego et al. 2013).

7.3.3.3 Last Millennium

Several high-resolution (bidecadal) records are accessible for the northern coasts of South America and nearby islands at the Colombian sites Ca, LF, and Hd, covering the past thousand years. These records consistently indicate a scarcity or absence of mangroves in earlier periods, with a conspicuous increase or appearance only in the recent centuries. LF site, for instance, witnessed a gradual expansion of mangroves between 450 and 100 cal yr BP (Castaño et al. 2010). In Ca, modern mangrove communities took root in the last century, having been virtually non-existent before then (Urrego et al. 2013). Conversely, Hd displayed a different pattern, with mangrove communities gradually expanding between 400 and 200 cal yr BP, reaching a peak and subsequently undergoing a sustained decline (González et al. 2010) (Fig. 7.12). These trends sharply contrast with the situation at another nearby site (NQ), where mangroves have experienced a millennial-scale decline since 2500–2000 cal yr BP and nearly disappeared in the last 500 years (Urrego et al. 2013). The spatial heterogeneity observed in this confined area (Fig. 7.4) is attributed to local variations, particularly in geomorphology, precipitation/evaporation rates, sediment input from rivers, and human disturbance. Addressing this heterogeneity is acknowledged as a significant challenge in developing a regional eustatic curve based on mangrove data (Urrego et al. 2013).

Among the climatic factors, the significant reduction in mangroves during the Little Ice Age (LIA) has been attributed primarily to severe droughts. These droughts likely led to hypersalinization, reduced sediment input, and coastal progradation, creating an unsuitable environment for mangrove development (Urrego et al. 2013). Mangrove communities may have also been impacted by extreme meteorological events such as hurricanes and destructive storms. For instance, at site Hd, an elevated phase of storm activity documented in historical records was identified as the cause of a sedimentary gap lasting over 2000 years before 1600 CE (González et al. 2010). In terms of human activities, widespread deforestation for various purposes and exploitative practices appears to have played a significant role. This is evident in the Hd pollen record, where a marked decline in *Rhizophora* by 1850 CE coincided with the rise of *Cocos nucifera* (coconut). The cultivation of coconuts, extensively practiced at the site and adjacent coasts since that time, is believed to be associated with this change (González et al. 2010; Urrego et al. 2019).

Moreover, along the Colombian coasts, the Ro record presented more comprehensive insights into the impact of climate on mangrove development over the past 400 years. The general finding was that climate played a role in shaping regional mangrove trends, while other factors had more localized effects. The record revealed that peaks in *Rhizophora* pollen followed periodic patterns linked to the Pacific Decadal Oscillation (PDO), the Atlantic Multidecadal Oscillation (AMO), and the El Niño-Southern Oscillation (ENSO) during the Little Ice Age (LIA) with 64-year cycles and between the mid-nineteenth and mid-twentieth centuries with 32-year cycles (Fig. 7.12). After 1954 CE, the *Rhizophora* record exhibited no significant cyclical patterns, likely attributed to escalating human disturbances, primarily associated with wood extraction, fisheries, coconut plantations, and rice crops. Although these mangroves demonstrated resilience to climatic cycles, their recovery rates were low in response to extreme or prolonged droughts and intense human activities (Urrego et al. 2019). Generally, factors such as extreme meteorological events (hurricanes, storms), coastal erosion or progradation, drought, salinization, and human disturbance have been identified as crucial drivers influencing mangrove dynamics in recent centuries. These drivers have resulted in either the expansion (progradation) or reduction (erosion) of mangroves, along with changes in their

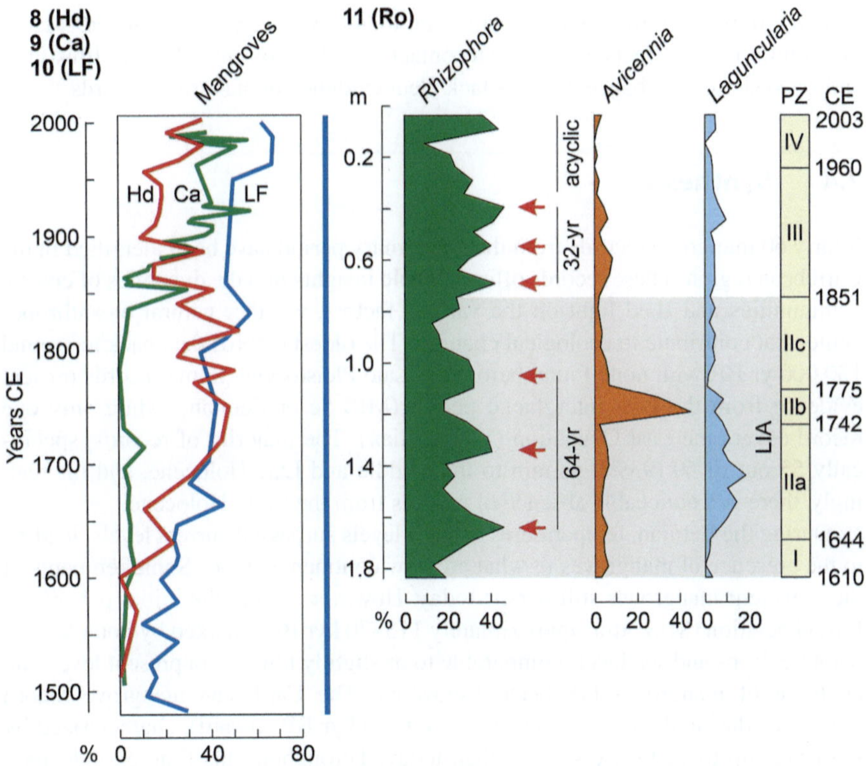

Fig. 7.12 Synthetic pollen trends illustrating mangrove dynamics on the Colombian Caribbean coasts during the last 500 years. See Fig. 19 for the locations of coring sites. PZ, pollen zones. Redrawn from Castaño et al. (2010), González et al. (2010), Urrego et al. (2013, 2019)

composition. Notably, *Laguncularia*, *Acrostichum*, and halophytic grasses replaced *Rhizophora* and *Avicennia*, which were removed by humans, even under high salinity conditions (Urrego et al. 2018).

In the Greater Antilles, mangrove communities in the Dominican Republic site Aj (Hispaniola) experienced significant impacts from hurricane activity over the past millennium. The MCA likely supported mangrove recovery after hurricanes due to relatively moist conditions, whereas the Little Ice Age (LIA), particularly after 330 cal yr BP, saw increased hurricane frequency and greater precipitation variability, causing delays in mangrove reestablishment. The higher charcoal influx during the MCA is attributed to increased fire incidence fueled by greater biomass availability in wetter conditions (LeBlanc et al. 2017).

Throughout the last millennium, notable cultural events have occurred, including the Maya civilization crises such as the Preclassic Abandonment (150–250 CE) and the Terminal Classic Collapse (750–950 CE). These events, linked to extreme droughts resulting from the migration of the ITCZ, do not leave distinct traces in existing mangrove records. Although some chronological correlations suggest potential impacts of these cultural shifts on mangroves, concrete evidence is lacking. Similarly, the effects of European contact on indigenous populations, lifestyles, and ecosystem disturbance patterns lack clear evidence in mangrove records.

7.4 Synthesis

Nearly 60 mangrove records from the Quaternary period have been identified in the Caribbean region. These records offer valuable insights into the dynamics of coastal communities and shed light on the various factors, whether natural or anthropogenic, that contribute to ecological changes. The oldest records date back to around 130,000 yr BP, with none found before the Late Pleistocene. Some records present evidence from the Last Interglacial period (MIS 5e or Eemian), while only one record covers the Last Glaciation (Weichselian). The majority of records, specifically 55 out of 60 (93%), pertain to the Middle and Late Holocene, and interestingly, there is a noticeable absence of records from the Early Holocene.

During the Eemian, temperatures and sea levels surpassed current levels, leading to the presence of mangroves in what are now continental areas. Some remnants of these ancient mangroves still persist today. However, during the initial part of the Last Glaciation (MIS 5d-a; approximately 115–70 kyr BP), marked by notable temperature drops and sea levels comparable to or slightly lower than present levels, no evidence of mangroves has been discovered. The Caribbean mangrove record reemerges during the coldest glacial phase (~70 kyr BP onward), characterized by sea levels up to 120 meters lower than today. Throughout this time period, mangroves underwent fluctuations in abundance in response to Heinrich stadials and Dansgaard–Oeschger interstadials, playing a minor role in coastal vegetation. Unfortunately, records for the Last Glacial Maximum (~21 kyr BP), the Late Glacial (up to 11.7 cal kyr BP), and the early Holocene (up to 8.2 cal kyr BP) are lacking, likely due to the shifting coastlines seaward and the submersion of potential sedimentary mangrove records under water columns ranging from approximately 20 to 120 m in depth. Holocene mangrove records become available from around ~8000 years ago when climates resembled those of today and sea levels were slightly lower. During this phase, mangroves primarily responded to the gradual eustatic rise characteristic of the Holocene.

Up until 6000–5500 cal yr BP, the ecological dynamics along the Caribbean coasts were predominantly shaped by natural factors. However, a notable shift occurred thereafter, with human influence becoming a significant factor. Human societies initially reached the Caribbean coasts through Central America approximately 13.5–12.5 cal kyr BP, eventually establishing themselves in the easternmost continental Caribbean region approximately 8 cal kyr BP. The Greater Antilles were settled from Central America by 7–6 cal kyr BP during the Lithic/Archaic Age, while the Lesser Antilles saw human arrival between 4 (Archaic Age) and 2 cal kyr BP (Ceramic Age). The earliest signs of human impact on Caribbean mangroves date back to Mesoamerica between 5600 and 5500 cal yr BP. During this period, Preclassic Maya societies practiced agriculture, specifically using fire to clear mangrove forests for cultivating maize and squash. This event, referred to as the Mesoamerican Archaic Disturbance (MAD), stands as one of the oldest records of agricultural activities in the region. In contrast, other areas in the Caribbean experienced relatively minor human influence on mangroves, with an exception noted in a site in the Orinoco Delta. It was not until the Late Holocene that evidence of human-induced mangrove disturbance became widespread across the Caribbean region.

The Late Holocene, which began 4200 cal yr BP, was marked by a significant regional climate shift known as the Cariaco Holocene Instability (CHI) event. During this period, there were notable fluctuations in precipitation and the occurrence of severe droughts. These climatic changes were likely influenced by the interaction between the latitudinal migrations of the Intertropical Convergence Zone (ITCZ) and increased activity of the El Niño-Southern Oscillation (ENSO). The CHI event played a crucial role in the decline of mangrove communities in various Caribbean areas. Arid climates during this period may have led to elevated salinity due to increased evaporation and reduced freshwater input from rivers. These conditions were unfavorable for the growth of mangroves. The possibility of climate-human amplification feedbacks during the CHI event has also been considered. The dynamics of mangrove ecosystems in the Late Holocene were found to depend on the balance between sediment transported from the continent and rising sea levels. When the equilibrium favored continental terrigenous input, coastal progradation countered the rise in sea levels, facilitating the seaward migration of mangrove communities. Conversely, in situations dominated by sea-level rise, mangroves tended to migrate landward. The Late Holocene witnessed diverse dynamics across the Caribbean region, emphasizing spatial heterogeneity in mangrove responses.

Over the last millennium, alterations in mangrove communities have primarily been driven by shifts in climate and human activities. The Little Ice Age (LIA), marked by cold and dry weather patterns, negatively impacted mangrove growth. Destructive weather events like hurricanes also played a notable role in diminishing mangrove communities. Notably, recovery after hurricanes occurred more swiftly during wet climatic phases, like the Medieval Climate Anomaly (MCA), and proved more challenging in the dry conditions of the LIA. In terms of human impact, historical and paleoecological records extensively document instances of mangrove deforestation for purposes such as wood extraction, fisheries, coconut plantations, and rice cultivation. The observed variations in mangrove changes are persistent and are linked to local factors such as coastal geomorphology, river dynamics, and human activities, while the influence of climate shifts tends to have a broader regional impact. Intriguingly, there is a lack of paleoecological evidence connecting mangrove disturbances to significant cultural events, such as the declines of the Maya Preclassic and Classic periods (around 200 CE and 800 CE, respectively) or European contact.

Future research endeavors should address certain knowledge gaps. For example, a notable void exists in the comprehensive Quaternary mangrove records, necessitating further investigative efforts. The absence of a complete Quaternary record poses a significant challenge, particularly in the absence of mangrove records spanning the entire Quaternary period. It is imperative to invest additional research efforts into this crucial area. The feasibility of acquiring a thorough Pleistocene-Holocene record appears to be most feasible in marine records. Notably, some Late Pleistocene pollen records discussed in this study were extracted from marine cores in the Venezuelan Cariaco Basin. These cores, retrieved during the Ocean Drilling Program (ODP), hold sequences dating back 600 kyr BP, offering a valuable resource for studying paleoclimatic and paleoceanographic proxies (Haug et al. 1998; Yarincik and Murray 2000). It is pertinent to investigate the remaining material in these cores for potential paleoecological studies on mangrove dynamics during the Pleistocene glacial-interglacial succession.

Furthermore, exploring the possibility of obtaining new complete Pleistocene sequences in the Venezuelan Cariaco Basin is of interest. A comprehensive Quaternary mangrove record for the Caribbean region is essential to augment the existing sequence, encompassing the Eocene origin and subsequent Oligocene and Neogene evolution of Neotropical mangrove communities (Chaps. 2, 3, 4, 5). Such an endeavor would contribute significantly to an integrated understanding of the ecological, biogeographical, and evolutionary history of Caribbean mangroves. In addition to shedding light on the Quaternary history of Caribbean mangroves, long marine records could be valuable for comparison with a unique continental record from central Colombia spanning the entire Pleistocene, which has been analyzed palynologically (Torres et al. 2013). Therefore, it is crucial to consider these avenues for comprehensive insights into the evolutionary trajectories and environmental dynamics of Caribbean mangrove ecosystems.

The exploration for additional Holocene mangrove records along the contemporary Caribbean coast is also needed and may benefit from the existing data discussed in this paper to identify more targeted research objectives. One potential avenue is to evaluate areas of recognized archaeological and historical significance, aiming to uncover records spanning the past millennium. This approach could shed light on the impact of recent cultural changes on mangrove development. A particular focus could be directed toward pivotal periods lacking sufficient evidence, such as the crises in Maya culture, the repercussions of European contact, or the responses to global warming over the last few centuries. Such an investigation would not only enhance our understanding of the formation of present mangrove communities but also contribute valuable insights for conservation efforts, particularly concerning potential irreversible thresholds related to climate change, sea-level rise, and anthropogenic stress.

References

Aburto-Oropeza O, Burelo-Campo CM, Ezcurra E et al (2021) Relict inland mangrove ecosystem reveals Last Interglacial sea levels. Proc Natl Acad Sci USA 118:e2024518118

Aragón-Moreno AA, Islebe GA, Roy PD et al (2018) Climate forcings on vegetation of the southern Yucatán Peninsula (Mexico) during the middle to late Holocene. Palaeogeogr Palaeoclimatol Palaeoecol 495:214–226

Baker PA, Fritz SC, Dick C et al (2014) The emerging field of geogenomics: constraining geological problems with genetic data. Earth-Sci Rev 135:38–47

Bocanegra-Ramírez DM, Li H-C, Domínguez G et al (2019) Holocene climate change and sea level oscillations in the pacific coast of Mexico. Quat Int 528:100–108

Boers N, Ghil M, Rousseau D-D (2018) Ocean circulation, ice shelf, and sea ice interactions explain Dansgaard-Oeschger cycles. Proc Natl Acad Sci USA 115:E11005–E11014

Bryan AL, Casamiquela RM, Cruxent JM et al (1978) A El Jobo mastodon kill at Taima-taima, Venezuela. Science 200:1275–1277

Bush MB, Silman MR, Listopad CMCS (2007) A regional study of Holocene climate change and human occupation in Peruvian Amazonia. J Biogeogr 34:1342–1356

Caffrey MA, Horn SP, Orvis KH et al (2015) Holocene environmental change at Laguna Saladilla, coastal north Hispaniola. Palaeogeogr Palaeoclimatol Palaeoecol 436:9–22

Carrillo-Bastos A, Islebe GA, Torrescano-Valle N (2013) 3800 years of quantitative precipitation reconstruction from the northwest Yucatan Peninsula. PLoS One 8:e84333

Castañeda-Posadas C, Correa-Metrio A, Escobar J et al (2022) Mid to late Holocene sea-level rise and precipitation variability recorded in the fringe mangroves of the Caribbean coast of Panama. Palaeogeogr Palaeoclimatol Palaeoecol 592:110918

Castaño A, Urrego L, Bernal G (2010) Dinámica del manglar en el complejo lagunar de Cispatá (Caribe colombiano) en los últimos 900 años. Rev Biol Trop 58:1347–1366

Castillo-Cárdenas MF, Ramirez-Silva JA, Sanjur O (2015) Evidence of incipient speciation in the Neotropical mangrove *Pelliciera rhizophorae* (Tetrameristaceae) as revealed by molecular, morphological, physiological and climatic characteristics. Bot J Linn Soc 179:499–510

Cerón-Souza I, Rivera-Ocasio E, Medina E et al (2010) Hybridization and introgression in New World mangroves, *Rhizophora* (Rhizophoraceae). Am J Bot 97:945–957

Cerón-Souza I, Bermingham E, McMillan WO et al (2012) Comparative genetic structure of two mangrove species in Caribbean and Pacific estuaries of Panama. BMC Ecol Evol 12:205

Cooke RG, Ranere AJ, Pearson GA et al (2013) Radiocarbon chronology of early human settlement of the Isthmus of Panama (13,000–7000 BP). Cultural affinities, environments and subsistence change. Quat Int 301:3–22

Coyne MK, Jones B, Ford D (2007) Highstands during Marine Isotope Stage 5: evidence from the ironshore Formarion of Grand Cayman, British West Indies. Quat Sci Rev 26:536–559

Creveling JR, Mitrovica JX, Clarck PU et al (2017) Predicted bounds on peak global mean sea level during marine isotope stages 5a and 5c. Quat Sci Rev 163:193–208

Davidson DE (2007) Modern pollen spectra from mangrove ecosystems of the sabana-camagüey archipelago and Ciego de Ávila, Cuba. MS diss, Univ Toronto, Canada

deMenocal PB, Oppo DW, Fairbanks RG (1992) Pleistocene d13C variability of North Atlantic intermediate water. Paleoceanography 7:229–250

DeYoe H, Lonard RI, Judd FW et al (2020) Biological flora of the tropical and subtropical intertidal zone: literature review for *Rhizophora mangle* L. J Coast Res 36:857–884

Dumas B, Hoang CT, Raffy J (2006) Record of MIS 5 sea-level highstands based on U/Th dated coral terraces of Haiti. Quat Int 145–146:106–118

Escarraga-Paredes DS, Torrescano-Valle N, Islebe GA (2014) Análisis de la relación vegetación-lluvia de polen actual de las comunidades vegetales en el nororoeste de la Península de Yucatán, Mexico. Polibotánica 38:27–52

Faegri K, Iversen J, Kaland E et al (1989) Textbook of pollen analysis. Blackburn Press, Caldwell

Fensterer C, Scholz D, Hoffmann D et al (2012) Cuban stalagmite suggests relationship between Caribbean precipitation and the Atlantic Multidecadal Oscillation during the past 1.3 ka. The Holocene 22:1405–1412

Fiedel SJ (2002) Initial human colonization of the Americas: an overview of issues and evidence. Radiocarbon 44:407–436

Fiedel SJ (2022) Initial human colonization of the Americas, redux. Radiocarbon 64:845–897

Fitzpatrick SM, Keegan WF (2007) Human impacts and adaptations in the Caribbean Islands: an historical ecology approach. Earth Environ Trans R Soc Edinburgh 98:29–45

Friedrich T, Timmermann A (2020) Using Late Pleistocene sea surface temperature reconstructions to constrain future greenhouse warming. Earth Planet Sci Lett 530:115911

González C, Dupont LM (2009) Tropical salt marsh succession as sea-level indicator during Heinrich events. Quat Sci Rev 28:939–946

González C, Dupont LM, Behling H et al (2009) Neotropical vegetation response to rapid climate changes duing the last glacial period: palynological evidence from the Cariaco Basin. Quat Res 69:217–230

González C, Urrego LE, Martínez JI et al (2010) Mangrove dynamics in the southwestern Caribbean since the 'Little Ice Age': a history of human and natural disturbances. The Holocene 20:849–861

Grauel AL, Hodell DA, Bernasconi SM (2016) Quantitative estimates of tropical temperature change in lowland Central America during the last 42 ka. Earth Planet Sci Lett 438:37–46

Hanagata N, Takemura T, Karube I et al (1999) Salt/water relationships in mangroves. Isr J Plant Sci 47:63–76

Haug GH, Pedersen TF, Sigman DM et al (1998) Glacial/Interglacial variations in production and nitrogen fixation in the Cariaco Basin during the last 580 kyr. Paleoceanography 13:427–432

Haug GH, Hughen KA, Sigman DM et al (2001) Southward migration of the Intertropical Convergence Zone through the Holocene. Science 293:1304–1308

Haug GH, Günther D, Peterson LC et al (2003) Climate and the collapse of Maya civilization. Science 299:1731–1735

Hearthy PJ, Hollin JT, Neumann AC et al (2007) Global sea-level fluctuations during the Last Interglaciation (MIS 5e). Quat Sci Rev 26:2090–2112

Hodell DA, Curtis JH, Jones GA et al (1991) Reconstruction of Caribbean climate change over the last 10,500 years. Nature 352:790–793

Hodell DA, Brenner M, Curtis JH (2005) Terminal Classic drought in the northern Maya lowlands inferred from multiple sediment cores in Lake Chichancanab (Mexico). Quat Sci Rev 24:1413–1427

Hofman CL, Antczak AT (2019) Early settlers for the insular Caribbean. Dearchaizing the Archaic. Sidestone Press, Leiden

Horn SP (1985) Preliminary pollen analysis of Quaternary sedimentsfrom Deep Sea Drilling Project site 565, western Costa Rica. Init Rep Deep Sea Drill Proj 84:533–547

Hughes PD, Gibbard PL, Ehler J (2013) Timing of glaciation during the last glacial cycle: Evaluating the concept of a global "Last Glacial Maximum" (LGM). Earth Sci Rev 125:171–198

Jackson ST, Williams JW (2004) Modern analogs in Quaternary paleoecology: here today, gone yesterday, gone tomorrow? Annu Rev Earth Planet Sci 32:495–537

Joo-Chang JC, Islebe GA, Torrescano-Valle N (2015) Mangrove history during middle- and late-Holocene in Pacific south-eastern Mexico. The Holocene 25:651–662

Keegan WF (2000) West Indian archaeology. 3, Ceramic Age. J Archaeol Res 8:135–167

Keegan WF, Hofman CL (2017) The Caribbean before Columbus. Oxford University Press, New York

Kennett DJ, Piperno DR, Jones JG et al (2010) Pre-pottery farmers on the Pacific coast of southern Mexico. J Archaeol Sci 37:3401–3411

Khan NS, Ashe E, Horton BP et al (2017) Drivers of Holocene sea-level change in the Caribbean. Quat Sci Rev 155:13–36

Koch MS, Snedaker SC (1997) Factors influencing *Rhizophora mangle* L. seedling development in Everglades carbonate soils. Aquat Bot 59:87–98

Lane CS, Clark JJ, Knudsen A et al (2013) Late-Holocene paleoenvironmental history of bioluminiscent Laguna Grande, Puerto Rico. Palaeogeogr Palaeoclimatol Palaeoecol 369:99–113

Lea DW, Park DK, Peterson LC et al (2003) Synchroneity of tropical and high-latitude Atlantic temperatures over the last glacial termination. Science 301:1361–1364

LeBlanc AR, Kennedy LM, Liu K-B et al (2017) Linking hurricane landfalls, precipitation variability, fires, and vegetation response over the past millennium from analysis of coastal lagoon sediments, southwestern Dominican Republic. J Paleolimnol 58:135–150

Leepart TP, Cochrane EE, Gaffney D et al (2022) Global patterns in island colonization during the Holocene. J Wolrd Prehist 35:163–232

Leyden BW (1985) Late Quaternary aridity and Holocene moisture fluctuations in the Lake Valencia basin, Venezuela. Ecology 66:1279–1295

Lonard RI, Judd FW, Summy KR et al (2017) The biological flora of coastal dunes and wetlands: *Avicennia germinans* (L.) L. J Coast Res 33:191–207

Lonard RI, Judd FW, DeYoe H et al (2020a) Biology and ecology of the halophyte *Laguncularia racemosa* (L.) Gaertn. f.: a review. In: Grigore M-N (ed) Handbook of halophytes. Cham, Springer Nature, pp 1–6

Lonard RI, Judd FW, DeYoe H et al (2020b) Biology of the mangal halophyte *Conocarpus erectus* L.: a review. In: Grigore MN (ed) Handbook of halophytes. Cham, Springer Nature, pp 1819–1831

MacIntyre IG, Toscano MA, Lighty RG et al (2004) Holocene history of the mangrove islands of Twin Cays, Belize, Central America. Atoll Res Bull 510:1–16

Martinez JI, Mora G, Barrows TT (2007) Paleoceangraphic conditions in the western Caribbean Sea for the last 560 kyr as inferred from planktonic foraminifera. Mar Micropaleontol 64:177–188

Martinson DG, Pisias NG, Hays JD et al (1987) Age dating and the orbital theory of the ice ages: development of a high-resolution 0 to 300,000-year chronology. Quat Res 27:1–29

McCloskey TA, Liu K-B (2012) A sedimentary-based history of hurricane strikes on the southern Caribbean coast of Nicaragua. Quat Res 78:454–464

McGirr R, Seton M, Williams S et al (2021) Kinematic and geodynamic evolution of the Isthmus of Panama region: implications for Central American Seaway closure. Bull Geol Soc Am 133:867–884

Mitchell SF, Pickerill RK, Stemann TA (2001) The Port Moran Formation (Upper Pleistocene, Jamaica): high resolution sedimentology and paleoenvironmental analysis of a mixed carbonate clastic lagoonal succession. Sedim Geol 144:291–306

Monacci NM, Meier-Grünhagen U, Finney BP et al (2009) Mangrove ecosystem changes during the Holocene at Spanish Lookout Cay, Belize. Palaeogeogr Palaeoclimatol Palaeoecol 280:37–46

Monacci NM, Meier-Grünhagen U, Finney BP et al (2011) Paleoecology of mangroves along the Sibun River, Belize. Quat Res 76:220–228

Montoya E, Pedra-Méndez J, García-Falcó E et al (2019) Long-term vegetation dynamics of a tropical megadelta: mid-Holocene palaeoecology of the Orinoco Delta (NE Venezuela). Quat Sci Rev 221:105874

Morrow JE, Morrow TA (1999) Geographic variation in fluted projectile points: a hemispheric perspective. Am Antiq 64:215–231

Muller J (1959) Palynology of recent Orinoco delta and shelf sediments. Micropaleontology 5:1–32

Nami HG (2021) Fishtailed projectile points in the Americas: remarks and hypotheses on the peopling of northern South America and beyond. Quat Int 578:47–72

Napolitano MF, DiNapoli R, Stone JH et al (2019) Reevaluating human colonization of the Caribbean using chronometric hygiene and Bayesian modeling. Sci Adv 5:eaar7806

Neff H, Pearsall DM, Jones JG et al (2006) Early Maya adaptive patterns: Mid-Late Holocene paleoenvironmental evidence from Pacific Guatemala. Latin Am Antiq 17:287–315

Nettel A, Dodd RS (2007) Drifting propagules and receeding swamps: genetic footprints of mangrove recolonization and dispersal along tropical coasts. Evolution 61:958–971

Nettel A, Dodd RS, Afzal-Rafii Z et al (2008) Genetic diversity enhanced by ancient introgression and secondary contact in East Pacific black mangroves. Mol Ecol 17:2680–2690

Newson LA, Wing ES (2004) On Land and Sea: Native American uses of biological resources in the West Indies. Univ Alabama Press, Tuscaloosa

Pearson GA (2002) Continental Paleoindian expansions and interactions as viewed from the earliest lithic industries of lower Central America. PhD Diss, Univ Kansas, Lawrence

Pearson GA (2017) Bridging the gap: an updated overivew of Clovis across Middle America and its techno-cultural relation with fluted point assemblages from South America. Paléo 3:203–230

Peterson LC, Haug GH, Hughen KA et al (2000) Rapid changes in the hydrologic cycle of the tropical Atlantic during the Last Glacial. Science 290:1947–1951

Pope KO, Pohl MED, Jones JG et al (2001) Origin and environmental setting of ancient agriculture in the lowlands of Mesoamerica. Science 292:1370–1373
Prates L, Ivan Perez S (2021) Late Pleistocene South American megagafunal extinctions associated with rise of Fishtail points and human population. Nat Commun 12:2175
Prufer KM, Alsgaard AV, Robinson M et al (2019) Linking late Paleoindian stone tool technologies and populations in North, Central and South America. PLoS One 14:e021981
Ramcharan EK, McAndrews JH (2006) Holocene development of coastal wetland at Maracas Bay, Trinidad, West Indies. J Coast Res 22:581–586
Ranere AJ, Cooke RG (2021) Late glacial and early Holocene migrations, and Middle Holocene settlement on the lower isthmian land-bridge. Quat Int 578:20–34
Ranere AJ, López CE (2007) Cultural diversity in the Late Pleistocene/Early Holocene populations in northwest South America and Lower Central America. Int J South Am Archaeol 1:25–31
Ravelo AC, Andreasen DH, Lyle M et al (2004) Regional climate shifts caused by gradual global cooling in the Pliocene epoch. Nature 429:263–267
Rinaldi M (1996) Palynological evidence of the Younger Dryas event in a core of Cariaco Basin, northeastern Venezuela (abstract). Proc 30th Int Geol Cong, Beijing, China, pp 4–14
Rodgers JC, Horn SP (1996) Modern pollen spectra from Costa Rica. Palaeogeogr Palaeoclimatol, Palaeoecol 124:53–71
Romano M (2015) Reviewing the term uniformitarianism in modern Earth Sciences. Earth-Sci Rev 148:65–76
Rouse I (1992) The Tainos: rise and decline of the people who Greeted Columbus. Yale University Press, New Haven
Rubio-Sandoval K, Rovere A, Cerrone C et al (2021) A review of last interglacial sea-level proxies in the western Atlantic and southwestern Caribbean, from Brazil to Honduras. Earth Syst Sci Data 13:4819–4845
Rull V (1998) Modern and Quaternary palynological studies in the Caribbean and Atlantic coasts of northern South America: a paleoecologically-oriented review. Bol Soc Venez Geól 23:5–24
Rull V, Vegas-Vilarrúbia T (1999) Surface palynology of a small coastal basin from Venezuela and potential paleoecological applications. Micropaleontology 45:365–393
Rull V, Vegas-Vilarrúbia T, Espinoza N (1999) Palynological record of an Early-Mid Holocene mangrove in eastern Venezuela. Implications for sea-level rise and disturbance history. J Coast Res 15:496–504
Rull V, Montoya E, Vegas-Vilarrúbia T et al (2015) New insights on palaeofires and savannisation in northern South America. Quat Sci Rev 122:158–165
Schmidt MW, Vautravers MJ, Spero HJ et al (2006) Western caribbean sea surface temperatures during the late Quaternary. Geochem Geophys Geosyst 7:Q02P10
Spratt RM, Liesecki LE (2016) A Late Pleistocene sea level stack. Clim Past 12:1079–1092
Stokes AV (2005) Ceramic-age dietary patterns in Puerto Rico: stable isotopes and island biogeography. In: Siegel PE (ed) Ancient Borinquen: archaeology and ethnohistory of Native Puerto Rico. Univ Alabama Press, Tuscaloosa, pp 185–201
Tamalavage AE, van Hengstum PJ, Louchouarn P et al (2020) Plant wax evidence for precipitation and vegetation change from a coastal sinkhole lake in the Bahamas spanning the last 3000 years. Org Geochem 150:104120
Tankersly KB, Dunning NP, Owen LA et al (2018) Geochronology and paleoenvironmental framework for the oldest archaeological site (7800–7900 cal BP) in the West Indies, Banwari Trace, Trinidad. Lat Am Antiq 29:681–695
Thompson WG, Goldstein SL (2006) A radiometric calibration of the SPECMAP timescale. Quat Sci Rev 25:3207–3215
Torres V, Hooghiemstra H, Lourens L et al (2013) Astronomical tuning of long pollen records reveals the dynamic history of montane biomes and lake levels in the tropical high Andes during the Quaternary. Quat Sci Rev 63:59–72
Urrego LE, Bernal G, Polanía J (2009) Comparison of pollen distribution patterns in surface sediments of a Colombian Caribbean mangrove with geomorphology and vegetation. Rev Palaeobot Palynol 156:358–375

Urrego LE, González C, Urán G et al (2010) Modern pollen rain in mangroves from San Andrés Island, Colombian Caribbean. Rev Palaeobot Palynol 162:168–182

Urrego LE, Correa-Metrio A, González C et al (2013) Contrasting responses of two Caribbean mangroves to sea-level rise in the Guajira Peninsula (Colombian Caribbean). Palaeogeogr Palaeoclimatol Palaeoecol 370:92–102

Urrego LE, Correa-Metrio A, González-Arango C (2018) Colombian Caribbean mangrove dynamics: anthropogenic and environmental drivers. Bol Soc Geol Mex 70:133–145

Urrego LE, Prado MA, Bernal G et al (2019) Mangrove responses to droughts since the little ice age in the Colombian Caribbean. Est Coast Shelf Res 230:196432

Van Daele M, van Welden A, Moernaut J et al (2011) Reconstruction of Late-Quaternary sea- and lake-level changes in a tectonically active marginal basin using seismic stratigraphy: The Gulf of Cariaco, NE Venezuela. Mar Geol 279:37–51

Van der Hammen TA (1963) palynological study on the Quaternary of British Guiana. Leidse Geol Mededel 29:125–168

Van der Stocken T, Carroll D, Menemenlis D et al (2019) Global-scale dispersal and connectivity in mangroves. Proc Natl Acad Sci USA 116:915–922

Vegas-Vilarrúbia T, Hernández E, Rull V et al (2015) The Orinoco megadelta as a conservation target in the face of the ongoing and future sea level rise. Sci Total Environ 515–516:129–142

Warken SF, Scholz D, Spötl C et al (2019) Caribbean hydroclimate and vegetation history across the last glacial period. Quat Sci Rev 218:75–90

Wooler MJ, Behling H, Leon J et al (2009) Late Holocene hydrologic and vegetation changes at Turneffe Atoll, Belize, compared with records from mainland Central America and Mexico. PALAIOS 24:650–656

Wurtzel JB, Black DE, Thunell RC et al (2013) Mechanisms of southern Caribbean SST variability over the last two millennia. Geophys Res Lett 40:5954–5958

Xia P, Meng X, Zhang Y et al (2021) The potential of mangrove-derived organic matter in sediments for tracing mangrove development during the Holocene. Est Coast 44:1020–1035

Yarincik KM, Murray RW (2000) Climatically sensitive aeolian and hemipelagic deposition in the Cariaco Basin, Venezuela, over the past 578,000 years: results from Al/Ti and K/Al. Paleoceanography 15:210–228

Rise and Fall: Summary and Conservation Insights

8

Abstract

The paleoecological and evolutionary findings discussed in this book are able to provide insights useful for conservation and restoration practices in the face of active deforestation, global warming, sea-level rise, or increased extreme events. The Caribbean mangroves have been structured around single mangrove-forming tree species—*Pelliciera* in the Eocene and *Rhizophora* from the Oligocene to the Pliocene—for most of their evolutionary history. This suggests that safeguarding the dominant species could ensure the continuity of mangroves as communities, while the preservation or restoration of other species could be a more gradual process. The heterogeneous responses of local mangrove communities to environmental shifts suggest that baseline conservation studies should be conducted on a site-specific basis. Regional initiatives are necessary to attract political, stakeholder, and public interest, but conservation/restoration actions should carefully consider local mangrove responses to site-specific threats for optimal results. For example, paleoecological studies demonstrate that it is essential to maintain a suitable balance between sediment supply, a local factor, and sea-level rise. Anthropogenic impact, also a regional factor with varied local manifestations, may be reduced by legislation, the creation of protected areas, and the promotion of restoration activities. In all cases, general regulations are needed, but local factors should also be considered.

This chapter emphasizes the importance of understanding the paleoecological and evolutionary background of ecosystems to inform biodiversity and ecosystem conservation. By examining past analogs and considering ecological responses to various environmental factors, such as paleogeographical, climatic, eustatic, and anthropogenic shifts, we can gain valuable insights. Knowledge of these biotic

V. Rull, *Origin and Evolution of Caribbean Mangroves*, Ecological Studies 252,
https://doi.org/10.1007/978-3-031-57612-6_8

responses becomes crucial in predicting potential future developments in ecosystems facing increasing natural and anthropogenic stress, in accordance with current global change estimates (IPCC 2022). The application of this knowledge can optimize conservation and restoration programs by drawing on past empirical evidence, as discussed by Willis et al. (2010) and extensively explored in the context of the Neotropics by Vegas-Vilarrúbia et al. (2011). This chapter specifically demonstrates the application of such a paleostrategy to mangroves, leveraging the comprehensive paleoecological and evolutionary information available in this book. By comparing the evolutionary and paleoecological history from the Paleogene, Neogene, and Quaternary intervals with the current state of Caribbean mangroves, including recent deforestation trends of both natural and anthropogenic origins, this chapter aims to extract lessons that can guide effective conservation and restoration efforts. A synthesis of Paleogene/Neogene evolutionary trends is represented in Fig. 8.1 to facilitate understanding.

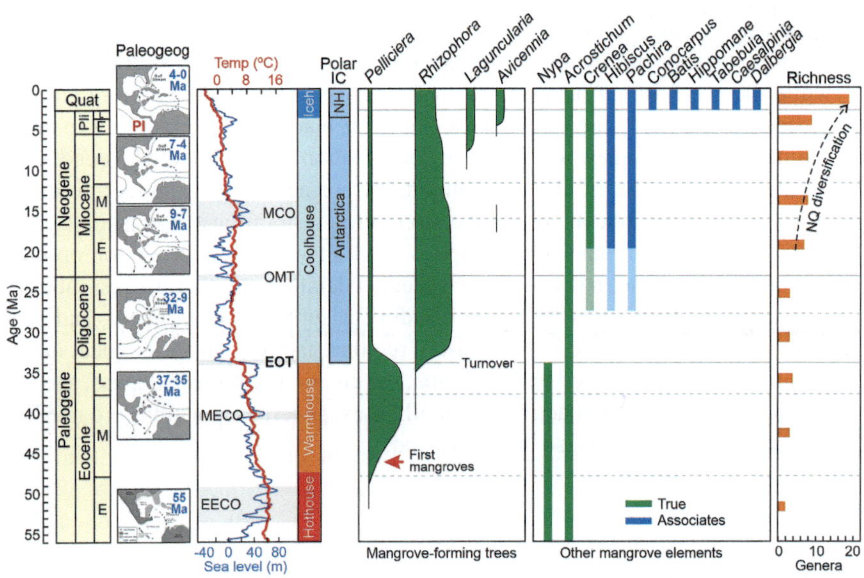

Fig. 8.1 Synthesis of the Paleogene/Neogene evolution of Caribbean mangroves in relation to paleographic (Iturralde-Vinent 2006), paleoclimatic (Westerhold et al. 2020), and paleoeustatic (Miller et al. 2020) shifts. Chronology: Quat, Quaternary; Pli, Pliocene; E, Early, M. Middle; L, Late. Paleogeography: PI, Panama Isthmus. Paleoclimates: EECO, Early Eocene Climatic Optimum; MECO, Middle Eocene Climatic Optimum; EOT, Eocene–Oligocene Transition; OMT, Oligocene/Miocene Transition; MCO, Miocene Climatic Optimum; Iceh, Icehouse; NQ, Neogene–Quaternary. Polar Ice Caps (IC): NH, Northern Hemisphere. Richness: NQ, Neogene-Quaternary

8.1 Evolution

8.1.1 Eocene

The conventional belief was that Neotropical mangroves originated from a global mangrove belt along the coasts of the Tethys Sea in the Late Cretaceous, over 65 Ma, following the formation of the African barrier due to continental drift. This notion was backed by qualitative evidence, including the presence of pollen and fossils from presumed mangrove elements like *Spinizonocolpites* and *Deltoidospora* (Srivastava and Prasad 2018). However, as elaborated in Chap. 3, a recent analysis of quantitative palynological evidence indicates that Neotropical mangroves likely emerged much later, between 50 and 40 Ma (Lutetian, Middle Eocene). Instead of being descendants of a former global belt, Caribbean mangroves are now considered an evolutionary innovation that arose independently. The earliest known mangrove-forming tree, *Pelliciera*, played a pivotal role in this evolution, dominating Eocene mangrove communities alongside *Nypa*, *Acrostichum*, and *Mauritia*. These novel communities rapidly spread across the region and were established in the Caribbean area by the Middle/Late Eocene. The updated CARMA dataset reveals a lack of evidence for mangrove communities before the Middle Eocene, and occasional occurrences of fossil pollen/spores from non-mangrove-forming elements, such as the palm *Nypa* or the fern *Acrostichum*, are deemed insufficient to substantiate the existence of mangrove forests.

8.1.2 Oligocene

During the transition from the Eocene to the Oligocene approximately 34 Ma, a noteworthy evolutionary shift occurred known as the Eocene–Oligocene transition (EOT). This period saw the replacement of ancient *Pelliciera* mangroves with the contemporary *Rhizophora* mangroves, as detailed in Chap. 4. *Rhizophora* was absent in the Neotropics during the Eocene and arrived in the Caribbean region during the EOT, likely through trans-Atlantic dispersal from its origin in the IWP region (Takayama et al. 2021). Pollen records indicate a sudden change in dominance from *Pelliciera* to *Rhizophora* during the EOT, coinciding with global cooling, a decrease in sea levels, and a significant biotic turnover, though not as catastrophic as the Big Five mass extinctions. The Eocene extinction and Oligocene radiation rates were heightened during this period (Coxall and Pearson 2007; Hutchinson et al. 2021). Notably, *Nypa* disappeared from the AEP region during the EOT. Interestingly, this major ecological shift did not result in the disappearance of *Pelliciera*. Instead, it became a minor component of the new *Rhizophora* mangroves since the Oligocene. *Pelliciera* extended its range to the entire Neotropics in the Miocene but persisted as a subordinate element, represented by small and scattered populations (Chap. 6). It is proposed that the continuity and further expansion of *Pelliciera*, a species with limited thermal tolerance and dispersal abilities, were facilitated by the protection provided by *Rhizophora*. This mangrove-forming tree, being more thermally tolerant and having better dispersal capabilities, created a microhabitat under its canopy, enabling *Pelliciera* to endure the new, less favorable environments arising from global disruptions during the EOT, akin to the dynamics observed today (Dangremond et al. 2015). Following the Miocene, *Pelliciera* underwent a significant reduction in its range, resembling its equatorial distribution in the Middle Eocene, leading to its present-day residual distribution.

8.1.3 Neogene

In the Neogene (Miocene–Pliocene), Caribbean mangroves experienced their main diversification trend. During this time, true mangrove elements such as *Crenea*, *Avicennia*, and *Laguncularia*, along with over 20 associated taxa, emerged, shaping the current richness patterns (Chap. 5). This led to an impressive 80% increase in diversity compared to the Paleogene (Eocene–Oligocene), where only half of the present-day true mangrove elements, such as *Rhizophora*, *Pelliciera*, and *Acrostichum*, were present. Notably, *Rhizophora* was the sole mangrove-forming tree during the Oligocene and much of the Neogene. *Laguncularia* and *Avicennia*, which are now important mangrove components, only became relevant in the Late Miocene and the Plio-Pleistocene, respectively. Importantly, no extinctions have been documented at the genus level in Caribbean mangroves since the Miocene, according to Graham (1995). The impact of climatic and sea-level fluctuations on the Neogene–Quaternary diversification trend (NQDT) remains uncertain.

8.2 Paleoecology

8.2.1 Pleistocene

For most of the Pleistocene (2.6 Ma to 11.7 cal kyr BP) in the Caribbean region, there is a notable gap in mangrove records, with the earliest evidence dating back approximately 130 cal kyr BP during the Eemian Interglacial, corresponding to MIS 5e, just before the Last Glaciation. The scarcity of data is attributed to the absence of complete Pleistocene records and emphasizes the necessity for extensive coring campaigns, particularly in marine environments (Chap. 7). During the Eemian interstadial in the Caribbean, sea surface temperatures (SSTs) were a few degrees higher than they are today, and it is estimated that sea levels were at least 3 m higher than the current levels, with a maximum estimate of 20 m. In contrast, during the Last Glacial Maximum (LGM) approximately 21 cal kyr BP, SSTs were 2–4 °C lower than present, and sea levels were as much as 120 m below their current position (Schmidt et al. 2006; Hearthy et al. 2007). Applying these patterns to past Pleistocene glacial–interglacial cycles suggests that comprehensive Pleistocene records should be sought in deep marine environments (>120 m depth), beyond the present continental shelf, where Caribbean coasts were likely located during glacial maxima. Existing records indicate that all current mangrove elements, both true and associated, were already in existence at the onset of the Pleistocene. This epoch was characterized by spatial and community reorganization driven by climatic and eustatic fluctuations. Nevertheless, additional research is required to confirm this hypothesis.

8.2.2 Holocene

In the Holocene (the last 11.7 cal kyr), a novel external environmental factor, human activity, emerged (Chap. 7). While Paleoindian settlements along the southern Caribbean coasts date back to approximately 13 cal kyr BP, significant disruptions to mangrove ecosystems did not occur until the Middle Holocene (~6 cal kyr BP). During this period, Mesoamerican Maya societies, particularly in present-day Mexico and Guatemala, began clearing mangrove forests using fire for maize and squash cultivation. Although these records are not strictly from the Caribbean region, they are included in the CARMA dataset due to their significance in the history of anthropogenic mangrove disturbance and their likely connection to the initial colonization of the Caribbean islands. Human colonization of the Greater and Lesser Antilles took place between approximately 6 and 2 cal kyr BP, with mangrove disturbance occurring afterward. Throughout the Holocene, human influence increased alongside rising temperatures and sea levels, as well as unpredictable moisture variations. Rising sea levels played a significant role in shaping mangrove communities, with their responses varying based on local conditions. The interaction between continental sediment input and sea-level rise was crucial for mangrove dynamics. When terrigenous input outweighed sea-level rise, coastal progradation prevailed, leading to seaward migration of mangrove communities. Conversely, dominance in sea-level rise favored landward migration. In the Late Holocene, regional changes in moisture levels and increased drought frequency/intensity became crucial factors, causing notable mangrove reductions due to salinity stress induced by elevated evaporation and reduced freshwater input from the continent (Rull 2022). Over the last millennia, human disturbances have escalated, as evidenced by paleoecological records documenting increased mangrove deforestation for wood extraction, fisheries, coconut plantations, and rice crops (González et al. 2010; Urrego et al. 2019).

8.3 Recent Decline and Future Projections

In recent years, the Caribbean mangrove area has experienced a significant decline due to both natural causes and human-driven deforestation. According to the latest information from the Global Mangrove Watch, the total mangrove coverage in this region decreased from 15,181 km^2 in 1996 to 14,677 km^2 in 2020, marking a 3.3% reduction over a span of 24 years (Bunting et al. 2022). This translates to an average annual loss of 21 km^2. While it is uncertain whether these deforestation rates will persist, the potential disappearance of Caribbean mangroves could take centuries, as outlined in Chap. 2. The impact of anthropogenic climate change, characterized by rising temperatures, increased aridification, and the resulting sea-level rise, poses an additional threat to Caribbean mangroves and those in the broader Neotropics (Farnsworth and Ellison 1996; Godoy and Lacerda 2015). Beyond outright removal, the fragmentation of mangrove communities constitutes an additional risk by increasing exposure to environmental stressors and diminishing their ability to provide vital ecological services, such as coastal protection and carbon sequestration (Bryan-Brown et al. 2020). However, protective measures and the changing dynamics of deforestation drivers with evolving economies could mitigate these challenges.

The current projections for global change in the Caribbean region anticipate a temperature rise of 1–3.5 °C by the end of the century, affecting both land and sea environments. Terrestrial settings are expected to experience the most significant warming, accompanied by a heightened occurrence of temperature extremes. Additionally, a moderate decrease in rainfall (20–30%) is likely, coupled with more frequent heavy rainfall events and prolonged hot and dry periods. The frequency and intensity of storms and hurricanes are also expected to increase based on global projections. Concerning sea levels, Caribbean forecasts coincide with global estimates, indicating a projected rise of 0.4–0.6 m relative to 1986–2005. Some studies even suggest the possibility of increases exceeding 1 m. These changes are anticipated to impact both terrestrial and marine ecosystems, posing risks to biodiversity. For instance, rising temperatures may lead to coral bleaching, the proliferation of invasive species, and the northward migration of Caribbean fish and coral reefs. Reduced rainfall could disrupt seed production, germination, and seedling development, potentially increasing seedling mortality. The heightened frequency of floods, storms, and hurricanes may accelerate the deterioration of coastal wetlands and forests, particularly mangroves, diminishing their natural filtering and buffering capacities. This degradation could result in habitat loss and contribute to the decline of ecosystems such as coral reefs, ultimately leading to biodiversity depletion. Additionally, sea-level rise may further impact mangroves by promoting their inland migration and exacerbating coastal erosion.

Hurricanes, also known as tropical cyclones, pose a significant threat to mangroves, particularly in the Caribbean region. Since 1995, there has been a notable rise in hurricane activity in this area (Webster et al. 2005; Palmieri et al. 2006; Burgess et al. 2018). Certain mangrove-rich locations in the Caribbean, such as Mexico and Cuba, are particularly susceptible to hurricanes, while others, such as South America, are infrequently impacted. The impact of tropical cyclones on mangrove ecosystems primarily manifests through increased tree mortality, alterations in forest structure and spatial distribution, and modifications to biogeochemical cycles (Krauss and Osland 2020). The influence of hurricanes on mangroves is contingent on various factors, including storm characteristics, geomorphic location, structural features of the affected forests, and regenerative capacity after the storm. The latter is connected to the dispersal potential of seeds and propagules, nutrient conditions, and ecophysiological features of the species involved. Despite the destructive nature of hurricanes, mangroves display a relatively high resilience, often undergoing regeneration. However, post-cyclonic secondary succession may lead to a different forest composition, contributing to increased spatial heterogeneity. The ongoing global changes in tropical cyclones are expected to result in more frequent and intense events, higher rainfall near the cyclone center, and a poleward shift in hurricane distribution. Consequently, the amplified effects of cyclones are likely to impact various regions, notably the Caribbean. Furthermore, anthropogenic activities causing fragmentation are expected to impede the regenerative capacity of mangroves (Krauss and Osland 2020).

It is worth noting that certain researchers contend that current predictions about future mangrove loss, stemming from both natural and human-induced factors, may be overly pessimistic. This viewpoint has given rise to a movement known as Conservation Optimism (Friess et al. 2020). These researchers posit that the anticipated changes in mangrove cover over time are prone to significant variability due to methodological issues. Additionally, they argue that rates of mangrove deforestation have actually decreased since the late twentieth century and assert that mangroves' responses to projected climate change can be anticipated. According to advocates of this optimistic perspective, these observations, coupled with a growing awareness of the vital role mangroves play in scientific, management, and policy contexts, have the potential to substantially enhance conservation and restoration efforts (Friess et al. 2019, 2022).

8.4 Conservation Insights

It is commonly asserted that what evolution has painstakingly developed over millions of years may be jeopardized within a few centuries. In the context of Caribbean mangroves, this evaluation can be quantified, considering that these ecosystems originated approximately 50 million years ago, and if current rates of loss persist, they could vanish in a matter of centuries. Recognizing the significance of mangroves for both terrestrial and marine biodiversity, ecology, and climate change mitigation, preserving these ecosystems has been deemed a priority in the Caribbean region. Numerous local and regional conservation and restoration initiatives have been implemented to address the situation, but they have proven insufficient to reverse the prevailing trend of mangrove loss and fragmentation (Barker 2002; Polidoro et al. 2010; Lacerda et al. 2019; Bryan-Brown et al. 2020; Walker et al. 2022). According to Lacerda et al. (2019), regional assessments and coordinated initiatives, especially involving the mangrove-bearing countries of Latin America and the Caribbean (LAC), are crucial for effective mangrove conservation and sustainable use. A regional approach is favored due to the diverse threats and impacts that vary significantly worldwide, making it challenging to adopt universal conservation policies (Ward et al. 2016). Within this framework, ecologically grounded knowledge is indispensable for establishing appropriate conservation and restoration benchmarks. Some of this knowledge can be gleaned from past paleoecological and evolutionary studies, such as those reviewed in this book. The following points could contribute to defining realistic conservation and restoration targets and ensuring their success.

8.4.1 Evolutionary Hints

For tens of millions of years, a single species of mangrove-forming tree, *Pelliciera*, dominated the Caribbean mangroves. During the Mid–Late Eocene, *Pelliciera* held sway, and in the Oligo-Miocene, *Rhizophora* took over as the dominant tree. It was not until the Pliocene and Late Miocene that the other structural mangrove trees, *Avicennia* and *Laguncularia*, made their appearance. This suggests that safeguarding the dominant species could ensure the continuity of mangroves as communities, while the preservation or restoration of other species could be a more gradual process. The fossil record teaches us that the components of ecological communities do not emerge all at once but evolve over time through innovative adaptations and progressive community assembly. New species integrate successfully into a community when the ecological conditions are favorable, highlighting the importance of allowing natural processes of community assembly in restoration practices, especially when the immediate reconstruction of an entire community is challenging or impractical.

Long-term expansions and contractions of geographical ranges can offer valuable insights for conservation purposes. A notable example is the case of *Pelliciera*, which thrived in ancient Eocene mangroves but is now confined to a small equatorial region as a highly endemic species. The historical range dynamics of this organism exemplify the taxon cycle, as proposed by Wilson (1961). This cycle involves an expansion from the original equatorial range (Eocene) to encompass the entire Neotropics (Miocene), followed by a subsequent contraction to its current distribution area. In line with taxon cycle predictions outlined in Chap. 7, *Pelliciera* may be approaching the final stages before facing natural extinction, a process that could be expedited by ongoing urban expansion (Blanco-Libreros and Ramírez-Ruiz 2021). Presently classified as "Vulnerable" in the IUCN Red List of Threatened Species (Polidoro et al. 2010), there is a suggestion that reevaluation in light of taxon cycle predictions could warrant its transfer to the "Critically Endangered" category.

8.4.2 Paleoecological Contributions

In Quaternary studies, another key factor highlighted is the equilibrium between sedimentation and sea-level fluctuations. This interplay governs coastal dynamics, influencing lateral movements of mangroves and ecological reorganizations in spatial terms. This aspect holds significant relevance today, given the observed rise in the Caribbean sea level at a rate of 1.8–2.5 mm per year (Palanisamy et al. 2012; Torres and Tsimplis 2013). Under these conditions, the effective conservation of mangroves in situ heavily relies on sedimentation, which is influenced by climate conditions and the characteristics of the local river system. Since climate factors are beyond human management, conservation efforts should concentrate on ensuring a consistent sediment supply capable of counteracting sea-level rise. This becomes particularly crucial in arid climates and regions with inadequate drainage, where engineering solutions may be necessary. Examining local records of mangrove colonization or degradation in response to changing sea levels can also provide valuable insights into the species involved at different successional stages, informing effective restoration practices.

Paleoecological and evolutionary research offers evidence-based historical parallels for evaluating the impact of climate change on mangrove ecosystems. This investigation can be conducted either on a regional or local scale. On a regional level, Caribbean mangroves have proven to be sensitive to prolonged fluctuations in temperature and moisture. However, this information may have limited applicability to specific conservation challenges, as it relies more on global policies to address ongoing anthropogenic global warming. On a local level, studies focused on specific mangrove communities and their reactions to climate variations are particularly enlightening. These studies provide precise, site-specific insights into how temperature and hydrological balance changes affect mangroves. Such responses can serve as valuable data for predicting potential future developments or can be integrated into broader predictive models at regional, continental, or global levels in the pursuit of identifying common patterns (Anderson et al. 2006). Chapter 7 contains numerous examples of individual case studies exploring how mangroves respond to shifts in climate and sea levels.

8.4.3 Anthropogenic Pressure

Paleoecological records commonly document the impact of human activities on ecosystems during the Middle–Late Holocene, particularly in recent centuries. The influence of human actions on Caribbean mangroves can manifest locally and directly through activities such as deforestation or more broadly and indirectly through factors such as global warming and sea-level rise. Effective conservation strategies encompass legislation, including the establishment of protected areas, as well as restoration efforts, with a focus on local interventions. Simultaneously, broader global policies aimed at mitigating climate change play a crucial role. Understanding the historical context of each specific site in terms of human interference can inform and enhance local assessments and actions. Meanwhile, global initiatives can draw insights from long-term paleoecological and evolutionary studies, as well as global paleoecological databases, to identify historical parallels and guide comprehensive conservation approaches.

Human activities contribute to environmental changes alongside natural forces, and at times, both factors interact synergistically. In paleoecological research addressing global change, a key objective is to untangle the influences of natural processes and human activities. This involves examining past records from times devoid of human impact, allowing for a focus on the dominance of natural forces such as paleogeographical, climatic, and eustatic drivers. The enduring presence of factors such as climatic and eustatic shifts after human impact can be achieved by correlating mangrove shifts with independent indicators of climatic and/or sea-level changes, specifically in the absence of anthropogenic markers such as cultivated plants and charcoal. Conversely, the prevalence of human-induced changes in mangroves can be inferred when anthropogenic proxies are present without significant environmental alterations. In instances where mangrove shifts align with evidence of both natural and human influences, the potential for synergistic interactions and amplification feedbacks should be considered. For instance, the combination of anthropogenic fires and arid climates may result in nonlinear responses, exacerbating forest loss.

In essence, the presence of human influence does not hinder the generation of valuable ecological information from historical records. Instead, it has the potential to help distinguish between the impacts of environmental factors and human activities. Furthermore, it can aid in identifying possible feedback loops and synergies, offering insights that are valuable for predicting future trends. Moreover, historical records, regardless of their type, provide evidence-based insights into long-term ecological processes that are not accessible through contemporary ecological studies.

References

Anderson HJ, Bugmann H, Dearing JA et al (2006) Linking palaeoenvironmental data and models to understand the past and to predict the future. Trends Ecol Evol 21:696–704

Barker DR (2002) Biodiversity conservation in the wider Caribbean region. Rev Eur Comp Int Environ Law 11:1

Blanco-Libreros JF, Ramírez-Ruiz K (2021) Threatened mangroves in the Anthropocene: Habitat fragmentation in urban coastalscapes of *Pelliciera* spp. (Tetrameristaceae) in northern South America. Front Mar Sci 8:670354

Bryan-Brown DN, Connolly RM, Richards DR et al (2020) Global trends in mangrove forest fragmentation. Sci Rep 10:7117

Bunting P, Rosenqvist A, Hilarides L et al (2022) Global Mangrove Watch: updated 2010 mangrove forest extent (v2.5). Remote Sens 10:1669

Burgess C, Taylor M, Spencer N et al (2018) Estimating damages from climate-related natural disasters for the Caribbean at 1.5 °C and 2 °C global warming above preindustrial levels. Reg Environ Chang 18:2297–2312

Coxall HK, Pearson PN (2007) The Eocene-Oligocene transition. In: Williams M, Haywood AM, Gregory FJ et al (eds) Deep-time perspectives on climate change: marrrying the signal from computer models and biological proxies. Geological Soc, London, pp 351–387

Dangremond EM, Feller IC, Sousa WP (2015) Environmental tolerances of rare and common mangroves along light and salinity gradients. Oecologia 179:1187–1198

Farnsworth EJ, Ellison AM, Gong WK (1996) Elevated CO2 alters anatomy, physiology, growth, and reproduction of red mangrove (*Rhizophora mangle* L.). Oecologia 108:599–609.

Friess DA, Rogers K, Lovelock CE et al (2019) The state of the world's mangrove forests: past, present, and futute. Annu Rev Environ Resour 44:89–115

Friess DA, Yando ES, Abuchahla GMO et al (2020) Mangroves give cause for conservation optimism, for now. Curr Biol 30:R153–R154

Friess DA, Adame MF, Adams JB et al (2022) Mangrove forests under climate change in a 2 °C world. WIREs Clim Change 13:e792

Godoy MDP, Lacerda LD (2015) Mangrove responses to climate change: a review of recent findings on mangrove extension and distribution. Ann Acad Bras Ciên 87:651–667

González C, Urrego LE, Martínez JI et al (2010) Mangrove dynamics in the southwestern Caribbean since the 'Little Ice Age': a history of human and natural disturbances. The Holocene 20:849–861

Graham A (1995) Diversification of Gulf/Caribbean mangrove communities through Cenozoic time. Biotropica 27:20–27
Hearthy PJ, Hollin JT, Neumann AC et al (2007) Global sea-level fluctuations during the Last Interglaciation (MIS 5e). Quat Sci Rev 26:2090–2112
Hutchinson DK, Coxall HK, Lunt DJ et al (2021) The Eocene-Oligocene transition: a review of marine and terrestrial proxy data, models and model-data comparisons. Clim Past 17:269–315
IPCC (2022) Climate change 2022: impacts, adaptation and vulnerability. Cambridge University Press, Cambridge
Iturralde-Vinent MA (2006) Meso-Cenozoic Caribbean paleogeography: implications for the historical biogeography of the region. Int Geol Rev 48:791–827
Krauss K, Osland MJ (2020) Tropical cyclones and the organization of mangrove forests: a review. Ann Bot 125:213–234
Lacerda LD, Borges R, Ferreira AC (2019) Neotropical mangroves: conservation and sustainable use in a scenario of global climatic change. Aquat Conserv 29:1347–1364
Miller KG, Browning JV, Schmelz WJ et al (2020) Cenozoic sea-level and cryospheric evolution from deep-sea geochemical and continental margin records. Sci Adv 6:aaz1346
Palanisamy H, Becker M, Meyssignac B et al (2012) Regional sea level change and variability in the Caribbean Sea since 1950. J Geodet Sci 2:125–133
Palmieri S, Teodonio L, Siani A-M et al (2006) Tropical storm impact in Central America. Meteorol Appl 13:21–28
Polidoro BA, Carpenter KE, Collins L et al (2010) The loss of species: mangrove extinction risk and geographic areas of global concern. PLoS One 5:e10095
Rull V (2022) Responses of Caribbean mangroves to Quaternary climatic, eustatic and anthropogenic drivers of ecological change: a review. Plan Theory 11:3502
Schmidt MW, Vautravers MJ, Spero HJ et al (2006) Western caribbean sea surface temperatures during the late Quaternary. Geochem Geophys Geosyst 7:Q02P10
Srivastava J, Prasad V (2018) Evolution and paleobiogeography of mangroves. Mar Ecol 40:e12571
Takayama K, Tateishi Y, Kajita T (2021) Global phylogeography of a pantropical mangrove genus *Rhizophora*. Sci Rep 11:7228
Torres RR, Tsimplis MN (2013) Sea-level trends and interannual variability in the Caribbean Sea. J Geophys Res 118:2934–2947
Urrego LE, Prado MA, Bernal G et al (2019) Mangrove responses to droughts since the little ice age in the Colombian Caribbean. Est Coast Shelf Res 230:196432
Vegas-Vilarrúbia T, Rull V, Montoya E et al (2011) Quaternary palaeoecology and nautre conservation: a general review with examples from the Neotropics. Quat Sci Rev 30:2361–2388
Walker JE, Ankersen T, Barchiesi S et al (2022) Governance and the mangrove commons: advancing the cross-scale, nested framework for the global conservation and wise use of mangroves. J Environ Manag 312:114823
Ward RD, Friess DA, Day RH et al (2016) Impacts of climate change on mangrove ecosystems: a region by region overview. Ecosyst Health Sustain 2:e01211
Webster PJ, Holland GJ, Curry JA et al (2005) Changes in tropical cyclone number, duration, and intensity in a warming environment. Science 309:1844–1846
Westerhold T, Marwan N, Drury AJ et al (2020) An astronomically dated record of earth's climate and its predictability over the last 66 million years. Science 369:1383–1387
Willis KJ, Bailey RM, Bhagwat SA (2010) Biodiversity baselines, thresholds and resilience: testing predictions and assumptions using palaeoecological data. Trends Ecol Evol 25:583–591
Wilson EO (1961) The nature of the taxon cycle in the Melanesian ant fauna. Am Nat 95:169–193

GPSR Compliance

The European Union's (EU) General Product Safety Regulation (GPSR) is a set of rules that requires consumer products to be safe and our obligations to ensure this.

If you have any concerns about our products, you can contact us on ProductSafety@springernature.com

In case Publisher is established outside the EU, the EU authorized representative is:

Springer Nature Customer Service Center GmbH
Europaplatz 3
69115 Heidelberg, Germany

Batch number: 10732044

Printed by Printforce, the Netherlands